The Ecological Risks of Engineered Crops

The Ecological Risks of Engineered Crops

Jane Rissler and Margaret Mellon

The MIT Press
Cambridge, Massachusetts
London, England

This book was set in Sabon by The MIT Press and printed and bound in the United States of America.

Library of Congress Cataloging-in-Publication Data

Rissler, Jane.
The ecological risks of engineered crops / Jane Rissler and Margaret Mellon.
p. cm.
Includes bibliographical references (p.) and index.
ISBN 0-262-18171-1 (hc : alk. paper). — ISBN 0-262-68085-8 (pb : alk. paper)
1. Transgenic plants. 2. Crops—Genetic engineering—Environmental aspects. 3. Plant genetic engineering—Environmental aspects. I. Mellon, Margaret G. II. Title.
SB123.57.R564 1996
631.5'23—dc20 95-38926
CIP

Contents

Foreword

Some of the most significant questions related to the future of the global economy and the global environment revolve around food and agriculture. How will it be possible to provide enough food for the ever-growing world population—a population annually adding the equivalent of another Mexico? Will global agricultural productivity decrease as a result of declining soil quality and depleted water resources? How can the environmental damage caused by pesticides and soil erosion be reduced? In the coming years, the Union of Concerned Scientists' Program on Agriculture and Biotechnology will evaluate some of the proposed solutions to the environmental and economic problems facing agriculture, not only in the United States but worldwide. We will advocate practices that will promote the long-term productivity and sustainability of agriculture, and that will reduce the risks of serious damage to people or the environment.

Private companies and the federal government are now pouring significant resources into biotechnology research, under the assumption that genetic engineering can solve many of the problems facing agriculture. Currently, the major application of genetic engineering to agriculture is transgenic crops—the focus of this study. This topic has extremely broad implications for the future of agriculture and the environment: are there serious risks associated with the large-scale release of transgenic crops into the environment?

This book considers some of the important and controversial issues surrounding the potential environmental risks of transgenic crops. Ideally, the discussion of risks would be embedded in a broader debate about the role of transgenic crops in the future of agriculture. This necessary debate would test the claims of dramatic benefits of transgenic crops and would consider other ways to improve agriculture.

As often happens, however, the momentum of commerce overtakes society's ability to fully discuss the risks and benefits of a new technology. Indeed, the first transgenic crops have already been commercialized in the United States and are currently moving into the international marketplace. The genie of agricultural biotechnology is out of the bottle and in many ways society is already playing catch-up.

We offer this book with the hope that it will produce two results. The first is to persuade the U.S. government to strengthen its regulatory programs and put them on a scientifically sound, experimental basis. The government needs a clear, orderly procedure for first deciding which transgenic plants are likely to be beneficial and which might have a long-term harmful effect on agriculture and the environment, and then for ensuring that those plants with likely harmful effects do not reach the market.

We also hope that our work will stimulate broader debate among policymakers and the public on the future role of transgenic crops. The paths available to transgenic crops are many. Even though the technology is out of the bottle, it is still early enough in its development that it can be channeled down several different paths. Which path will be taken depends on whether agendas are set by private industry or public research institutions, whether products are directed to the needs of the industrialized or the developing world, and whether scientists, farmers, and consumers participate in the debate on an equal footing with industry and government. The issues are complex and people of goodwill differ on the answers. But we believe that there is still need for a serious public discussion on the extent to which the various applications of crop biotechnology are likely to facilitate or retard progress toward a more sustainable agriculture in the United States and abroad.

Howard Ris
Executive Director
Union of Concerned Scientists

Preface

Recent technical feats involving living organisms have launched us on a long journey. We are not very far down the road. But the breakthrough has been made. Scientists can now purposefully and efficiently create genetic novelty by transferring and rearranging genes and groups of genes.

Just several years ago, few people imagined that cows might be used as pharmaceutical factories—a development already upon us. No one can imagine the products of the genetic engineering industry of 2025. Because the products of nature and classical breeding are constrained by mechanisms of natural reproduction, changes come slowly and thus appear to depart little from the familiar. We are moving, however, into a world where the look, behavior, and uses of organisms will have few bounds—and where we shall have few guideposts. Much current uneasiness about the technology derives from the sense of having started down a path that could lead in unwanted directions.

This book focuses on one aspect of the effects of genetic engineering: the environmental risks associated with commercial-scale use of engineered, or transgenic, crops. Transgenic crops are a subset of the large number of living organisms that are being genetically engineered by agricultural, pharmaceutical, food processing, and other industries for a range of purposes from food and fiber production to toxic waste degradation. Most transgenic crops will not harm the environment; some will undoubtedly be beneficial.

We focus on the risks of bioengineered crops for three reasons. First, many transgenic crops are being commercialized and will be planted in large acreages in this country. Second, some transgenic crops may cause environmental problems, including effects on centers of biological diversity both here and abroad. And third, the federal government's regulatory program needs to be strengthened to adequately manage the environmental risks. This book describes and categorizes those risks and offers a general scheme to assess them in the context of a government oversight program. The risk assessment scheme includes an innovative approach to predicting environmental risks based on empirical observations of transgenic plant behavior.

The environmental risks of transgenic crops are not generic; that is, the plants do not share any one particular trait, such as fast growth or aggressiveness. Instead, their risks depend on the novel combinations of traits that modern gene transfer technologies make possible. Most genetically engineered organisms will not be harmful. At the same time, the overall likelihood of harm will rise as the number and variety of crop releases increase. The risks are similar in some ways to those of the introductions of nonnative organisms into new environments. Most nonnative organisms die out quickly in new environments. But, occasionally one will take hold, and—in the absence of ecological controls—thrive to cause extensive damage.

Unfortunately, the long-term, cumulative risks to ecosystems of introducing large numbers of transgenic plants are not well enough understood to allow their prediction except in the grossest sense. It is unlikely that ecosystem dynamics will be well enough understood any time in the near future to confidently predict this aspect of environmental impact. Obviously, unknown risks cannot be anticipated or evaluated.

Our tentative conclusion, at this early stage in the development of the technology, is that unregulated releases of transgenic plants pose serious risks requiring a scientifically sound regulatory system. The conclusion is informed by humankind's long experience with traditional plant breeding as well as by the fragmentary nature of current understanding in the fields of ecology and genetics. But our view is open to modification as society gains greater experience with transgenic plants in the environment and as scientists deepen their collective understanding of the principles of ecology and genetics.

There are several kinds of potential risks associated with the commercialization of transgenic crops. First, the transgenic crops themselves may become weeds. Second, they may serve as a conduit through which new genes move to wild plants, which could then become weeds and require expensive control programs. The novel genes may also affect wild ecosystems in ways that are difficult to evaluate. Third, plants engineered to contain virus particles may facilitate the creation of new viruses. New viral pathogens that affect economically important crops could require significant control costs. Fourth, plants engineered to express potentially toxic substances like drugs and pesticides could present risks to other organisms that are not the intended targets of the new chemicals. For example, drug-producing plants could poison birds feeding in corn fields. Fifth, we also consider the cascading effects that can ripple through an ecosystem as the result of an initial perturbation, and address the fact that some risks of transgenic crops may not yet be known.

Finally, we address the pressing global problem that the commercialization of transgenic crops could threaten global centers of crop diversity. Populations of wild plants and landraces (traditional varieties), growing in proximity to one another over a long, uninterrupted span of time, result in a great concentration of crop biodiversity. These centers of diversity, located primarily in the developing world, are the natural reservoirs for the traits needed to maintain the vitality of modern crops. Among the most challenging risks of transgenic crops is the threat that they might pose to these important regions of diversity, upon which U.S. and world agriculture depends.

There is broad agreement among governments that steps must be taken to minimize the health and environmental risks of transgenic organisms that come to market. We believe action is needed now to strengthen the U.S. regulatory scheme to accomplish this goal. As our society focuses on risks and regulations, however, we should not lose sight of the broader debate about this technology. The advent of engineered crops raises profoundly important social and economic questions about the control of the world's food supply, the need to adopt environmentally sustainable agricultural practices, and the pressing demands of feeding the burgeoning human population. It is important to emphasize that, so far, there is no consensus that transgenic crops are essential to the solution of problems of agricultural sustainability or world food production. Uncertainty about risks requires society to consider related benefits as well. Society should be willing to assume more risks for applications of transgenic plants that will feed hungry people, for instance, than for those that would merely increase a chemical company's market share for a herbicide.

Proponents of biotechnology often assume, almost automatically, that the development of transgenic crops should be the primary response to the challenges of twenty-first century agriculture. Others, however, question how essential many applications of biotechnology will be to increasing global food supply and reducing the environmental problems associated with farming. They see crop rotations, diversity in crops and animals, reduced dependence on synthetic inputs, and other practices associated with the concept of "sustainable agriculture" as the best starting point for making agriculture less environmentally damaging while maintaining high productivity. In their view, biotechnology is most useful when it furthers these practices rather than reinforces conventional, industrial agriculture.

These are complex issues that deserve searching public debate. Society must acknowledge that genetic engineering presents new terrain; it is difficult to know the extent to which past experience with traditional breeding is a valid predictor of the risks this new technology poses. The complexity of ecological processes, and humankind's incomplete knowledge of how they operate, lead us to believe that sufficient resources must be devoted to risk assessment of such crops before they are commercialized and that decisionmakers should err on the side of caution in the face of uncertainty.

Acknowledgments

We gratefully acknowledge the support of the Joyce Foundation, whose grant to the National Wildlife Federation (NWF) Biotechnology Policy Center made this project possible. Joyce Foundation funds supported all phases of the project, from the 1991 experts' workshop held under the auspices of NWF to the final writing, production, and distribution of the report that led up to this book, undertaken at the Union of Concerned Scientists. This book is a revised, expanded version of the earlier report: *Perils Amidst the Promise: Ecological Risks of Transgenic Crops in a Global Market.*

The experts' workshop on transgenic plants that laid the foundation for chapters 3 and 4 was held in 1991 in Washington, D.C. We are grateful to workshop participants not only for their thoughtful deliberations during the discussions, but also for their subsequent reviews of draft manuscripts. Workshop participants included: Norman Ellstrand, Department of Botany and Plant Sciences, University of California, Riverside; James Hancock, Department of Horticulture, Michigan State University; Kathleen Keeler, School of Biological Sciences, University of Nebraska, Lincoln; C. Randal Linder, Program of Ecology and Evolutionary Biology, Brown University (currently at the Department of Biology, Indiana University); Robin Manasse, Department of Zoology, University of Washington (currently with the Environmental Research Laboratory, Environmental Protection Agency, Corvallis, Oregon); Philip Regal, Department of Ecology and Behavioral Biology, University of Minnesota; Mary Lynn Roush, Forest Science Department, Oregon State University.

We are particularly grateful to Robin Manasse, who contributed significantly to the risk-assessment proposal offered in chapter 4. Dr.

Manasse contributed to both the evolution of the fundamental approach and its implementation in the suggested testing scheme.

In addition, we note the contributions of four scientists—Candace Collmer, Gus de Zoeten, Peter Palukaitis, and Thomas Pirone—who reviewed the portion of the book dealing with virus-resistant plants, an issue not addressed in detail at the original workshop.

We are grateful to Michael Bayliss, Edward Bruggemann, Kelly Donegon, Michael Hansen, Peter Kareiva, Hope Shand, Bernice Slutsky, Nachama Wilker, and Hugh Wilson, who reviewed the manuscript at various stages in its preparation.

Among the many other people who contributed to the production of this book, we extend special thanks to Jan Wager Reiss and Seth Shulman for editing, Sharon Levy for finding important resource materials, Holly Haynes for assisting in the preparation of the tables and charts, and Tunisia R. Dunn for preparing the final manuscript.

Although we appreciate the valuable advice and information provided by the workshop participants and other reviewers, this book does not necessarily reflect their opinions. Participation in the workshop or willingness to review drafts should not imply an endorsement of the book or its recommendations. The authors assume full responsibility for any errors.

1

Introduction

The risks associated with transgenic plants exist against a backdrop of enormous changes in the biological sciences. Genetic engineering is no longer simply a research tool. Advanced techniques are poised to become the basis of new industries with the potential to transform major sectors of human society and the environment. Among other things, genetic engineering offers the increased domestication of forests and fisheries, the replacement of food plants and animals with cell cultures, and expanded, nonagricultural uses of plants and animals. Each of these developments represents a choice, not an inevitability.

At the core of both the promise and the risks of genetic engineering is the power to produce combinations of genes not found in nature. Modern gene transfer techniques allow scientists to directly transfer functional genetic material to host organisms. Human genes can be transferred to cows, plant genes can be transferred to bacteria. These techniques vastly increase the ability to generate organisms with new properties.

Scientists have already produced cows that serve as drug-production facilities, secreting human proteins in their milk for use as pharmaceuticals; they have engineered plants with genes that sequester metals so they can be fertilized with metal-contaminated sludge; and, by inserting the genes of fireflies, scientists have created tobacco plants that glow for use as test systems for light-based organismal markers. Several research groups are engineering crops to produce vaccines against human diseases. Some of the work is devoted to developing edible vaccines—transgenic vaccine-producing crops which can be eaten as part of a normal diet thereby immunizing the consumers against diseases (Moffat, 1995).

What interests people most about genetic engineering, however, is not its present but its future. Since, theoretically, functional genes can be taken from any organism and transferred into any other organism, the potential reach of the technology is very broad, indeed.

At present, numerous impediments reduce the range of practical gene transfers to far less than the theoretical limit. For example, only a tiny fraction of the genes in nature have been identified and isolated; existing transfer techniques limit potential hosts; the techniques are restricted to small blocks of deoxyribonucleic acid (DNA) that usually contain only one or two genes of interest; the final locations of transgenes[1] on host chromosomes are random, so insertions of new sequences may interfere with the function of existing genes; and organisms have functional limits to their tolerance of the disruptive effects of added genes. But most of these restrictions are technical in nature and will not impede the technology for long. The library of genes available for transfer grows daily. At the same time, the list of possible host organisms is increasing as scientists develop new techniques for gene transfer, such as the bioballistic technique that propels genes into cells on tiny metal projectiles (Sanford et al., 1987). Finally, as the understanding of gene function and regulation continues to improve, progress can be anticipated in both the number of genes per transfer and the availability of mechanisms to avoid disruptive effects.

In the realm of agricultural biotechnology, small-scale field tests of genetically engineered crops (transgenics) have been under way in the United States for almost eight years now, and a number of products are on the market or nearing the point at which they could be commercialized. Leading the pack was Calgene, a California biotechnology company that received approval to market the first genetically engineered whole food—a tomato—in 1994. Since then several other crops, including insecticidal potatoes, virus-resistant squash, and high lauric acid canola, have been approved. Many other companies have transgenic products that are poised to follow the first wave into the marketplace.

Many important issues surrounding commercialization of transgenic crops remain unsettled. Among these are the impacts of such products on the sufficiency and viability of world agriculture. There is increasing concern about the ability of international food production systems to feed an exploding world population in the twenty-first century, and to do so in a way that will not cause massive environmental harm. Many see the solution to these complex problems in a fundamental reorientation of agriculture toward sustainable practices that apply a systems-based, rather than a single-component approach, to the perpetual problems of yield, pest control, and soil conservation.[2] Although largely beyond the scope of this book, an important question for the future is whether engineered crops will facilitate or retard a global transition to a sustainable agriculture.

As these crops come to market, there are pressing concerns about the environmental risks entailed in the wide commercial use of transgenic crops. This issue has received some discussion in government and industry circles, but often from a narrow perspective that has downplayed the seriousness of the risks. This book attempts to bring the environmental risks of engineered crops to center stage.

The subject matter covered here is limited in an important way. We discuss only the environmental issues raised by the commercial use of transgenic plants, and do not address most of the human and animal health issues. Analyzing even this restricted set of issues is a formidable challenge. If the agricultural biotechnology industry fulfills the hopes of its promoters, it could be producing hundreds of kinds of transgenic vegetables, grains, fruits, trees, fiber crops, and ornamentals by the turn of the century. These plants will be grown on vast acreage in the United States and around the world. Most of these crops will contain novel combinations of genes and traits. And in many cases, these genes will be transferred via pollen from the crops to populations of wild relatives.

What environmental risks will large numbers of these different varieties of transgenic crops pose? It is currently impossible to provide a complete answer to this question. Ecological risks depend, among other things, on the nature of the crop, the characteristics of the added gene, and the agricultural locale. As a modest first step, we offer a framework with which to begin to analyze these environmental risks. It includes a practical, innovative approach for assessing some of the ecological risks for purposes of decisionmaking in a regulatory program.

Genetic Engineering versus Traditional Breeding

Both natural mutation processes and traditional breeding produce organisms with new genetic traits. In general, these activities have received little government oversight because experience has shown that the behavior of plants produced by traditional breeding is predictable and only occasionally troublesome. Why are genetically engineered crops eliciting a higher level of concern than traditionally bred crops?

There are no properties or risks attached to an organism just because it has been engineered. Instead, risks arise from the explicit properties of the added genes, the effects of new combinations of genes, and specific environmental situations. In particular, because ecological effects emerge from complex interactions of organisms, they are often unexpected and difficult to predict. Even in retrospect, for example, it is difficult to explain why kudzu has been so successful at finding and exploiting an ecological niche, while so many other introductions of exotic plants have failed.

Genetic engineering is obviously different and more powerful than traditional breeding. The question is whether this difference and power translate into greater, more serious risks, or whether the risks will simply be different from those of traditionally bred products. At this early stage in the use of the technology, it is impossible to answer this question with certainty. But there are a number of reasons to be concerned.

First, since genetic engineers have an enormous pool of new genes to draw on, they can add more genes with harmful potential than can traditional breeders. In contrast to gene splicing, traditional breeding is limited to the gene pool of plants that can sexually interbreed with a crop. Ecological risk depends on complex and difficult-to-predict interactions between new genes, the crops into which they are spliced, and the environment in which the new crop is grown. Generally, the greater the variety of new genes that can be added to a crop via genetic engineering, the greater the likelihood of something going awry.

Second, the novelty and power of the technology suggest that its outcome may be less predictable than the results of traditional breeding.[3] Genetic engineering can add or subtract complete genes or reconstitute genes in novel ways. Traditional breeding, on the other hand, is often limited to substituting slightly different versions of genes already present—

not new genes—in the crop or wild relatives. Unlike the replacement of one version of a gene for another, a wholly new gene may interact with the rest of the plant genome in more unpredictable ways.

The idea that gene splicing may bring with it greater unpredictability is supported by observations suggesting that genes transferred from distantly related organisms often may be expressed under fewer constraints in recipient plants than genes from more closely related organisms (Ninio, 1983; Regal, 1988). Relevant to this concept is work showing that gene modifiers in bacteria evolve to alter the expression of a new mutation (Lenski, 1988a, b). In transgenic organisms, this may mean that modifiers of the transgene will not have time to evolve—resulting in less control over the expression of the new gene.

Third, many of the transgenes that are being moved into crops control traits that appear obviously advantageous to plants. These new traits, such as resistance to disease or insects, may enable weeds to overcome ecological limits on population growth. Most of the transgenic crops field tested so far have been given ecologically advantageous traits. By contrast, many of the traits established in most crops by traditional breeding tend to reduce rather than enhance the fitness of wild relatives (Ellstrand, 1988; Regal, 1992). Traditional breeding often selects for versions of genes that are agronomically important but interfere with natural adaptability. For example, the rapid germination bred into some plants is advantageous to irrigated crops but detrimental to wild relatives that might germinate quickly after a spring rain only to die well before a second rain.

Lowered fitness is not necessarily significant in the survival of a crop because it is so intensively managed by humans. Farmers compensate by using management techniques that substitute for many traits that would lower fitness in nature. The new, lower-fitness versions of genes added in traditional breeding are unlikely to be useful to a wild population and would not be retained.

Fourth, the capacity to combine genes from dramatically different organisms has, at least until recently, little known precedent in evolution. There is not yet a large body of knowledge and experience on which to predict outcomes. By contrast, traditional breeding is usually limited to crosses among closely related organisms—within species and almost always within genera. In other words, genes are combined among organisms whose genetic makeups are similar.

Finally, transgenic crops used as "factories" for pharmaceuticals and specialty chemicals present a unique challenge to agricultural and nonfarm ecosystems. Soil insects and microorganisms, foraging and burrowing mammals, seed-eating birds, and a myriad of other nontarget organisms will be exposed for the first time to vaccines, drugs, detergent enzymes, and other chemicals expressed in the engineered plants. Herbivores will consume the chemicals as they feed on plants. Soil microbes, insects, and worms will be exposed as they degrade plant debris. Aquatic organisms will confront the drugs and chemicals washed into streams, lakes, and rivers from fields.

Scientists have little knowledge of the macroenvironmental consequences of producing transgenic plants. Given enough time and a broad enough selection of engineered crops, movement of transgenes into wild relatives of crop plants is a virtual certainty. Thus, the widespread adoption of genetically engineered crops will mean a constant flow of novel genes not only into agricultural ecosystems, but through such systems into wild ecosystems as well. Many of these genes will come from animals and bacteria and would not ever have found their way into plants without genetic engineering. What will it mean to have a steady stream of animal and microbial genes entering the gene pools of plants in wild ecosystems? It may mean little. But, as yet, it is clear that we lack even a framework within which to answer the question.

A New Empirical Approach to Risk Assessment

Much of this book is based on the deliberations of a workshop of scientists assembled in the fall of 1991 to identify the risks of commercial-scale use of engineered crops and, where possible, develop a scheme for assessing those risks. Other parts of the book grew out of discussions that ensued with the exchange of drafts after the workshop was over. These discussions eventually led us to add a section on the risks of virus-resistant plants.

In addition, these discussions resulted in the development, in conjunction with several workshop participants, of a new approach to risk assessment that relies far more strongly on field observations than the original workshop scheme did. The new empirical approach provides a practical, feasible method of conducting risk assessments that will assist governments seeking to identify and protect against two of the environmental risks of transgenics.

We hope that the empirical approach to risk assessment offered here will be seriously considered by government agencies charged with regulating transgenic plants. Currently, these agencies rely heavily on armchair analyses of how plants should behave, rather than on observations of how they do behave. Although the scheme proposed here also has its limitations, it offers a novel approach that may be more practical and protective than existing ones.

Road Map to the Text

Chapter 2 defines transgenic plants and reviews research currently underway in the field of agricultural biotechnology. The chapter offers a brief overview of the environmental risks posed by transgenic crops as well as a discussion of why current field tests tell little about potential risks from their commercial-scale production.

Chapter 3 identifies and categorizes the environmental risks presented by commercial uses of transgenic crops. First, transgenic crops may become weeds, either on the farm or in wildlife habitats; the term "weed" is used broadly to include any unwanted plant. The chapter discusses several aspects of weediness. The first category focuses on the initial impact of a new weed that may persist in a farmer's field or invade a wild habitat. The second category of risk is that novel genes will flow from the transgenic crop to wild relatives and create new weeds. Third, we discuss the possibility that virus-resistant transgenic crops could produce new viruses or alter the host range of existing viruses.

Chapter 4 proposes a three-tiered scheme for evaluating whether a transgenic crop will become a weed. The assessment relies primarily on data generated on the transgenic crop itself. Similarly, a three-tiered approach is offered to assess the capacity for transgene flow to create new weeds in wild plants that are relatives of the crop. Finally, this chapter explains the difficulty of assessing other aspects of weediness.

Chapter 5 addresses the international implications of widespread adoption of transgenic crops. Of particular concern is the flow of novel genes into cultivated and wild relatives that constitute centers of diversity for crops. The flow of new genes into these centers, most of which are in developing countries, could in the future threaten the genetic base of the world's food supply.

Chapter 6 reviews the current status of U.S. regulations governing transgenic crops, discusses the difficulties that unknown risks present to regulators, and makes the case for widespread public involvement in the issue. Finally, this chapter outlines the Union of Concerned Scientists' conclusions and recommendations concerning the commercial development, risk assessment, and regulation of transgenic crops.

2 Understanding Transgenic Crops

During its first twenty years, biotechnology was largely an indoor activity. Its major applications in biomedical research and drug manufacture involved mainly the use of microorganisms on laboratory benches and in fermentation vats. But biotechnology has now begun to leave the laboratory and enter the environment. After a decade of development, chemical and seed companies are beginning to commercialize the first transgenic crops. The commercialization of transgenic crops is important for the environment because it will mean the release of many genetically engineered organisms under uncontrolled conditions. As commercialization proceeds, environmental risks, which have until now been largely hypothetical, will become real.

What Are Transgenic Plants?

Transgenic plants are crops that have been genetically engineered to contain traits from unrelated organisms. Genetic engineering refers to sophisticated, artificial techniques capable of transferring genes from other organisms directly to recipient organisms. Genes determine specific traits, like color, height, or tolerance to frost. Adding novel genes to crops means adding new traits and abilities. Genetic engineers can move genes from any biological source—animals, plants, or bacteria—into almost any crop.

To give a simple example, a traditional breeder interested in producing a yellow tomato must find the yellow trait in a plant that will breed with the tomato by natural mechanisms. The only plants that can breed with tomatoes are closely related ones. Unrelated plants like oak trees or

Table 2.1
Sources of new genes in transgenic crops

Crop	Source of new genes	Purpose of engineering
Potato	Chicken	Increased disease resistance
	Giant silk moth	Increased disease resistance
	Greater waxmoth	Reduced bruising damage
	Virus	Increased disease resistance
	Bacteria	Herbicide tolerance
Corn	Wheat	Reduced insect damage
	Firefly	Introduction of marker genes
	Bacteria	Herbicide tolerance
Tomato	Flounder	Reduced freezing damage
	Virus	Increased disease resistance
	Bacteria	Reduced insect damage
Tobacco	Chinese hamster	Increased sterol production
Rice	Bean, Pea	New storage proteins
	Bacteria	Reduced insect damage
Melon, Cucumber, Squash	Virus	Increased disease resistance
Sunflower	Brazil nut	Introduction of new storage proteins
Alfalfa	Bacteria	Production of oral vaccine against cholera
Lettuce, Cucumber	Tobacco Petunia	Increased disease resistance

Source: Information compiled from applications to the U.S. Department of Agriculture to field test engineered plants (Union of Concerned Scientists, 1993).

cantaloupes could not breed with tomatoes and thus could not contribute new genes. A genetic engineer, on the other hand, can consider any organism—even a butterfly or a daffodil—as a source of the yellow trait. If the gene that determines yellow color has been identified and isolated, it can be directly transferred into tomato plants. Thus, transgenic crops can contain combinations of genes and traits that traditional breeding could not produce. In that sense, transgenics are a genuinely new class of plants, or crops.

Scientists have used the recombinant technology to insert genes from plants, animals, bacteria, and viruses into a host of crop plants (table 2.1). For example, genes from chickens and silk moths have been spliced into potatoes to confer resistance to bacterial diseases. One group at the Illinois-based Amoco Technology Company inserted genes from a Chinese hamster into tobacco to increase the crop's production of sterol—an unsaturated solid alcohol product. Because sterol has industrial and agricultural applications, the research could lead to the use of tobacco plants as sterol-producing factories. Another company, the California firm, DNA Plant Technology, added antifreeze protein genes from winter flounder into tomatoes. The research sought to create a tomato that can withstand colder temperatures.

Transgenic Plants Are Being Commercialized

After a slow start, the agricultural biotechnology industry is proceeding rapidly toward the marketplace. Scientists have succeeded in producing engineered versions of most of the world's major food and fiber crops—including corn, rice, potato, soybean, and cotton—as well as numerous fruits, vegetables, and trees (table 2.2).

Many of these engineered plants have moved from the laboratory to the field-testing stage. Since 1987, the U.S. government has approved hundreds of applications to field test engineered plants, with many crops now in their fourth or fifth year of testing. Worldwide, there have been well over 1,500 approvals, many covering more than one field-test site. In the United States, the first genetically engineered crop offered for commercial sale was a delayed-ripening tomato created by the California-based biotechnology firm Calgene. The tomato cleared all regulatory hurdles in the summer of 1994. Later that year, a transgenic virus-resistant squash from Asgrow Seed Company, a subsidiary of Upjohn Company, became the second engineered crop approved for commercial sale.[1] Since then a number of other transgenic crops, including insecticidal potatoes, lauric acid-producing canola, and a second delayed-ripening tomato have been approved. A number of crops will follow these onto market shelves.

To date, based on reported field-test activity in the United States, chemical companies are the major players in agricultural biotechnology (table 2.3). Seed companies, land grant universities, and the U.S. Department of Agriculture (USDA) are also performing research and

Table 2.2
Transgenic plants under development

Cereal	Corn, Rice, Rye
Fiber	Cotton, Flax
Forage	Alfalfa, Orchard grass
Forest	Poplar, Spruce
Fruit/nut	Apple, Cranberry, Grape, Kiwi, Melon, Papaya, Pear, Plum, Strawberry, Walnut
Oil	Peanut, Rapeseed (canola), Soybean, Sunflower
Ornamental	Allegheny serviceberry, Carnation, Chrysanthemum, Morning glory, Petunia, Rose
Vegetable	Asparagus, Cabbage, Carrot, Cauliflower, Celery, Cucumber, Eggplant, Horseradish, Lettuce, Pea, Potato, Squash, Sweet potato
Miscellaneous	*Arabidopsis*, Chicory, Foxglove (*Digitalis*), Licorice, Sugar beet, Tobacco

Sources: Fraley, 1992; Union of Concerned Scientists, 1993.

conducting tests on transgenic crops. In some cases, important financial ties exist between universities and corporations in the biotechnology area (Krimsky et al., 1991; Wisconsin Rural Development Center, 1993). In addition, a few small biotechnology companies in the United States, most in collaboration with large corporations, are working to develop agricultural crops.

A Realistic Look at Plant Biotechnology

Early proponents of plant biotechnology billed it as a revolutionary technology—one that would easily produce miracle crops that would resist drought, fix their own nitrogen from the air, and resist disease. They saw a bright future for agriculture: corn thriving without added nitrogen fertilizer; cotton plants withstanding insects without added

Table 2.3
Applicants to the U.S. Department of Agriculture to field test transgenic crops[a]

Applicants	Percent of applications
Chemical companies	46
Monsanto[b]	
Upjohn (Asgrow Seed)	
DuPont	
Sandoz (Northrup King and Rogers NK Seed)	
Ciba-Geigy	
Hoechst-Roussel	
Imperial Chemical Industries	
American Cyanamid	
Universities/U.S. Department of Agriculture	17
USDA Agricultural Research Service	
Cornell University	
North Carolina State University	
University of Kentucky	
University of California	
Michigan State University	
Seed companies	15
Pioneer Hi-Bred[c]	
DeKalb Plant Genetics	
Holden's Foundation Seed	
Petoseed	
Harris Moran	
Biotechnology companies (stand alone)	13
Calgene[d]	
DNA Plant Technology	
Agrigenetics	
Food companies	5
Frito-Lay	
Campbell	
Heinz	
Land O'Lakes	
Miscellaneous	4
Cargill	
Amoco Technology	

a. Information compiled from 549 applications submitted to the U.S. Department of Agriculture (from 1987 to summer 1993) to field test transgenic crops (Union of Concerned Scientists, 1993). In each category, only the most active participants are listed.
b. Monsanto accounts for 52 percent of the applications in this category.
c. Pioneer Hi-Bred accounts for 51 percent of the applications in this category.
d. Calgene accounts for 82 percent of the applications in this category.

insecticides; drought-tolerant crops reclaiming environmentally degraded land. With these and other transgenic plants, it seemed world agriculture could stave off the looming global food crisis. All this and a high return on investments—these were the dreams of plant biotechnology in the early 1980s.

After ten years in the laboratory and the field, plant genetic engineering has proved to be more difficult, taken longer, and cost more than the early enthusiasts imagined (*Economist*, 1995). Genetic engineers discovered that they were limited in terms of the number of genes—two to five—that can be transferred and the kinds of traits that can be modified with existing, even though growing, gene libraries. These proved to be important road blocks. The most desirable traits—yield, drought resistance, nitrogen fixation—are determined by many genes. Nitrogen fixation, for example, requires a complex interplay of dozens of plant and bacterial genes (Hirsch, 1992). For traits like yield and drought-resistance, scientists are only beginning to understand the full complement of genes that determine the trait.

These limits and other difficulties have led the biotechnology industry to sharply scale back its expectations for plant genetic engineering—at least for the foreseeable future. Increased yield, drought resistance, salt tolerance, and other abiotic environmental stress resistances have been set aside in favor of alterations with less sweeping implications. Indeed, scientists and companies working on transgenic crops no longer mention ideas like nitrogen-fixing corn.

Below is a brief discussion of some of the first wave of crops produced by agricultural biotechnology.

Herbicide-Tolerant Crops

Currently, the major application of plant biotechnology is to make plants tolerant to specific chemical herbicides. The products offer new weed-killing options to farmers in the form of herbicides that could not be used on nonengineered varieties.

Most of these crops are being developed by, or in conjunction with, chemical companies who also sell the herbicides. Over 40 percent of the applications to field-test genetically engineered crops in the United States are for herbicide-tolerant crops (Union of Concerned Scientists, 1994).

Table 2.4
Transgenic crop traits tested in industrialized countries, 1986–1992
A study by the Organization for Economic Cooperation and Development[a] shows the percentage of small-scale field trials approved for six transgenic-crop traits in fourteen member countries[b] from 1986 to 1992.

Trait	Percent[c]
Herbicide tolerance	57
Virus resistance	13
Insect resistance	10
Quality traits[d]	8
Male sterility	5
Disease resistance	4

a. "Evaluation of Biosafety Information Gathered During Field Releases of GMO's" [DSTI/STP/BS (92)6], summarized in Friends of the Earth European Coordination, 1993.
b. The countries are Australia, Belgium, Canada, Denmark, France, Germany, the Netherlands, Norway, Japan, Spain, Sweden, Switzerland, the United Kingdom, and the United States.
c. Percentages do not add to 100 because this list contains only the six most-tested, agronomically important traits. Not included are traits that were tested only a few times and tests of plants engineered to contain biological markers.
d. Quality traits include altered fruit ripening, altered oil or protein composition of seeds, and increased solids content of fruits and tubers.

In the industrialized world, herbicide-tolerant crops account for 57 percent of the approvals for field-testing transgenic crops (table 2.4).

There is considerable debate over the impact of engineered herbicide tolerance on the use of chemicals in agriculture. Proponents of herbicide-tolerant crops acknowledge that such crops will require continued use of chemicals, but they argue that there will still be important environmental benefits. They believe that farmers could use fewer applications of herbicides in crops that otherwise might require multiple applications over a growing season. They also claim that engineered crops will permit farmers to substitute newer "environmentally benign" herbicides for older more toxic ones. Those who hold a contrary view believe that the development of herbicide-resistant crops will simply prolong agriculture's dependence on hazardous chemical inputs.

The development of bromoxynil-tolerant cotton, the first transgenic herbicide-tolerant crop approved for commercial use, provides a case in point. On the one hand, the Environmental Protection Agency considers bromoxynil to be a developmental toxin and a possible human carcinogen (Environmental Protection Agency, 1992, 1995); it is also highly toxic to fish (Hunn et al., 1989). Thus, anything that would facilitate and encourage the use of bromoxynil raises important questions about potential adverse impacts. However, if the use of transgenic cotton would allow the number of herbicide applications to be reduced and if bromoxynil is found to be less toxic than the herbicides now being used, the net result could be beneficial. The key issues are whether the newer herbicides for which tolerance is sought will be less hazardous than those they replace and whether their application will actually reduce herbicide use.

Pest-Resistant Crops

After herbicide tolerance, the most popular application of genetic engineering is the production of crops that resist pests such as insects, viruses, and fungi. Insect and fungal resistance are especially desirable because the pesticides used to combat insects and fungi can be highly toxic.

Different molecular strategies are being employed to combat different pests. In the case of insects, the added genes produce insect toxins; in the case of viruses, the new gene products interfere with viral multiplication. Virus- and insect-resistant crops have been approved for commercialization. Research seeking genes for fungal resistance is under way, but such crops are only in the early stages of development.

Pest resistance is a generally desirable application of transgenic crops. Keeping pests at bay increases yields and, in some cases, allows a reduction in the application of pesticides, thereby lowering costs and reducing environmental damage. Unfortunately, the benefit of the initial transgenic crops in repulsing pests may be short-lived. Most of the first round of insect-resistant crops, for example, employ the same (or very similar) toxin gene(s) to produce insect resistance in crops including corn, rice, cotton, tobacco, and many others. If a number of these products were to be offered for sale, the same toxin, called the Bt toxin, would soon be produced at high levels in a wide variety of crops. This is a recipe for the development of resistance in insect pests. Where resistance to the toxin occurs, the genetically engineered plants will lose their effectiveness, and conventional farmers may once again have to resort to chemicals. Moreover, resistance would mean that organic and other low-input farmers, who now rely on Bt toxin (sold in the form of the soil bacterium that produces it naturally), could lose one of their most valuable pest-control agents.

Genetic engineering could be used to retard the development of resistance to biological pesticides. One approach would be to engineer a crop so that a toxin is only expressed if the crop is under attack by the target pest. Bt versions of such plants are now under development but none are among the Bt crops currently proposed for commercialization. Whether the concept will prove to be successful and can be commercialized in time to influence the development of Bt resistance remains to be seen.

Processing Traits and Longer Shelf Life

Processing and transport applications were rarely included in the early list of revolutionary applications of biotechnology. They lack glamour because they appeal primarily to those involved in the intermediate steps in the food system. But according to a recent report by the Biotechnology Industry Organization (BIO) on the future of agricultural biotechnology, much of genetic engineering, like much of traditional breeding, will be devoted to such applications (BIO, 1995). Examples of crops engineered for processing traits are high-solids tomatoes currently under development by Monsanto (BIO, 1995). Such tomatoes mean that less water must be removed in the making of salsa and ketchup, a feature that would clearly be advantageous to manufacturers. In a similar vein, Monsanto is engineering a high-solids potato that will absorb less oil, reduce cooking time, and lower costs to processors, and DNA Plant Technology is adding genes to control the texture of frozen strawberries (BIO, 1995).

Products being engineered for longer shelf life include tomatoes (one said to last up to forty days after harvest), pineapples, bananas, and cherry tomatoes (BIO, 1995). Longer shelf life increases flexibility and lowers costs for those who transport and sell fresh produce. In some cases, these improvements could lead to lower prices for consumers.

Consumer Niche Markets

Agricultural biotechnology is also producing premium products for consumer niche markets. In fact, crop biotechnology's commercial debut was a better-tasting supermarket tomato. Supermarket tomatoes are usually tasteless because flavor has been sacrificed in favor of transportability. Calgene introduced a tomato engineered with a gene that delayed ripening, enabling the tomato to stay for a longer time on the vine without softening. The idea was that flavor would develop in tomatoes still tough enough to survive the rough going between field and market. DNA Plant Technology is marketing a delayed-ripening tomato engineered by a different process and is also headed for niche markets with delayed-ripening cherry tomatoes, sweeter snap peas, and firmer peppers (BIO, 1995).

To date, improving the nutritional value of food does not appear to have received as much emphasis as shelf life and processing traits. Only one such product is included in the recent BIO report, a tomato with higher levels of anti-oxidant vitamins (BIO, 1995).

Summary

In summary, the current products of plant biotechnology do not represent a revolutionary breakthrough. If current trends continue, the output of biotechnology will resemble that of traditional breeding. It will be aimed primarily at growers, processors, and transporters, with a smaller set of premium products aimed at consumer niche markets. Use of some of the products might lead to reduction in the use of pesticides, although such reductions are in no way guaranteed. As herbicide-tolerant crops demonstrate, transgenic crops can be produced as companions to rather than replacements for chemical inputs.

Transgenic crops will eventually join their traditionally bred counterparts in the market, sometimes displacing them, sometimes not. Whether the sum of this mixed bag of products will be considered beneficial or not will depend on a number of factors, including one's view of today's conventional agricultural systems. Biotechnology fits comfortably into modern food systems that emphasize food processing, consumer niche markets, and production efficiency. For those interested in environmentally sound agriculture, biotechnology may offer incremental improvements, but it may also fall short of the fundamental reform that so excited its early proponents.

Feeding the World

In the industrialized world, agriculture is already highly productive and most of the people are well, if not over, fed. Hunger in these societies has nothing to do with production shortages. But what about the developing world? Might not genetic engineering be important to parts of the world where agricultural production lags, often the same parts of the world facing the brunt of the expected increases in global population? Can biotechnology, as is often claimed, feed the world?

World hunger is a complex and pressing problem that deserves a far more thorough analysis than can be given here. But there are a number of reasons to suspect that here, too, the contributions of transgenic plants will be limited.

First, transgenic crops are largely being developed by corporations that are focusing, as all profit-making entities must, on paying customers. As a result, virtually all transgenic crops—including corn, soybean, and tomatoes—are aimed at the prosperous farmers of the North. Comparatively little work involves cassava, plantain, or cowpea, which millions in the developing world depend on for home consumption and local markets. Companies occasionally give away engineered versions of crops to farmers in developing countries, but such gestures cannot produce the broad variety of locally adapted crops needed to make a difference.

Second, genetic engineering may not offer significant technical advantages over traditional breeding for increasing levels of crop production. Traditional plant breeding is a powerful and proven technology that has contributed significantly to the abundance of agricultural production that the industrialized, and much of the developing, world currently enjoys. If resources for traditional breeding in developing countries were on a par with the research and extension resources available in the industrialized world, there is little doubt that dramatic increases in food production would ensue. Whether genetic engineering could increase productivity further or faster than traditional breeding is an important, but still open, question. The agricultural biotechnology products discussed above do not demonstrate any such advantage, but the technology is young and this question needs to be revisited as the technology matures.

Third, even if research is done on the right crops and results in increased agricultural productivity, increased production is only one factor in the complicated equation of world hunger. Poverty, trade policies, subsidies, soil erosion, and water shortages are also important causes of hunger. Increases in productivity obtained through genetic engineering will have mixed effects if not developed with due regard to the other important aspects of the hunger problem.

If properly directed, transgenic crops might make a positive contribution to increasing global food production. Delineating a role for agricultural biotechnology is among the many challenges facing society if we are to meet the looming global food crisis. At the outset of this debate, however, it is important to qualify the expectation that transgenic crops—or biotechnology as a whole—can somehow single-handedly solve the problem of world hunger. High-technology crops will not serve as a technological fix for hunger. They will not compensate for decades of environmental abuse, misguided agricultural policies, and income disparities. It would be tragic if reliance on biotechnology lulled decision-makers into complacency on this vital issue.

Biotechnology and the Future of Agriculture

The Union of Concerned Scientists has no moral or ethical objection to the use of sophisticated gene transfer technology per se. Nor do we presume that the glitziest technology necessarily offers the best solution to problems. Instead, we take a pragmatic middle-of-the-road approach that looks at each application of biotechnology on its own merits, considering its risks and benefits, and its alternatives.

In general, the research applications of genetic engineering have made possible giant strides in the understanding of the biological world and how it operates. Also, many of the pharmaceutical applications of engineered bacteria offer effective therapies to patients, pose few risks to those who do not derive benefits, and have few viable alternatives. In agriculture, however, the situation is murkier. Here the benefits of agricultural biotechnology to a country awash in food and agricultural commodities are less obvious, the risks of environmental harm are greater, and the alternatives are more plentiful.

Agricultural biotechnology should be evaluated in the context of efforts to make agriculture more sustainable and less environmentally damaging. Conventional intensive agricultural practices, including reliance on monoculture—planting the same crop year after year—and synthetic inputs, have damaged the environment, contaminated groundwater, eroded topsoil, and threatened the health of farmers and farm workers. Proponents of biotechnology have viewed the technology as a way to reduce dependence on inputs such as pesticides and fertilizers. But transgenic crops are not the only way to reduce chemical use. Advocates of "sustainable agriculture" take an ecological approach that adjusts the parameters of agroecosystems with sophisticated practices and information strategies, often obviating rather than "solving" problems (National Research Council, 1989a). They emphasize healthy soil and promote such practices as diversifying crops and animals, rotating crops, and intercropping. They support research that studies soil, water, climate, crops, animals, pests, and wildlife on a farm as an interrelated whole.

The advantages of this ecological, systems-based approach can be seen in pest control. Sustainable farmers do not focus on pests one at a time but instead take steps to prevent problems from arising in the first place. They start by building healthy soil and choosing plant varieties resistant to pests. They plant different crops in the same field in successive years, thereby reducing pests that may build up on any one crop. Providing habitats for other organisms—from birds to parasitic wasps—can also help keep down the levels of pests.

In contrast, monoculture as practiced by industrial agriculture inevitably gives insect pests opportunities to build up their numbers with a continuous supply of food. The conventional approach then views the resulting pests as an isolated problem and tries to control them with a pesticide. If the pests develop resistance to the pesticide, the farmer turns to another pesticide. Conventional agriculture therefore requires a large arsenal of poisons. Most current applications of biotechnology accept monoculture as a given and try to solve pest problems by equipping each crop with new insecticidal genes.

Crop rotations, along with other sustainable agriculture practices, can produce crops decade after decade without yield loss, water pollution, or other degradation of resources. Of course, rotations are not panaceas.

Once they are in place, there is still a need for some pest-control measures. Transgenic crops might be useful to control residual pests or to make rotations work. But the commitment to rotation must come first if genetic engineering is to offer truly compatible products. Focusing on transgenic crops first could produce marginal improvements in monoculture systems but never get to the root of the problem—monoculture itself. In addition, too much enthusiasm for "miracle" plants could divert resources needed to support the transition to an agriculture built on sustainable principles.

The Environmental Risks of Transgenic Crops

Genetically engineered crops are not inherently dangerous; they only present problems where the new traits, or combinations of traits, made possible by modern gene technology produce unwanted effects in the environment. Different genetically engineered crops will present different problems depending on the new genes they contain, the characteristics of the parent crop, and the locale in which they are grown.

Because the number of crops and genes is so large and varied, identifying and categorizing potential risks of transgenic crops is a challenge. To keep this analysis from becoming unwieldy, we have limited its scope by focusing on the set of issues that arise primarily as a result of plant interactions with other organisms in the environment. We do not address the food safety issues associated with the consumption of genetically engineered food or feed. Nor do we address the toxicological impacts of engineered crops except to the extent that consumption of toxic plants might alter ecosystem compositions. It does address two major categories of environmental risks: those of the engineered plants themselves and those associated with the movement of transgenes into other plants (figure 2.1).

Broadly speaking, the risks from the engineered plants themselves are that the new traits might enable them either to become weeds in agricultural ecosystems or to move out of the field to disturb unmanaged ecosystems. A crop that is engineered to resist insect pests, for example, may become better able to survive in unmanaged ecosystems; in most situations, this event may not matter because the newly advantaged plants

1. The transgenic crops themselves become weeds.

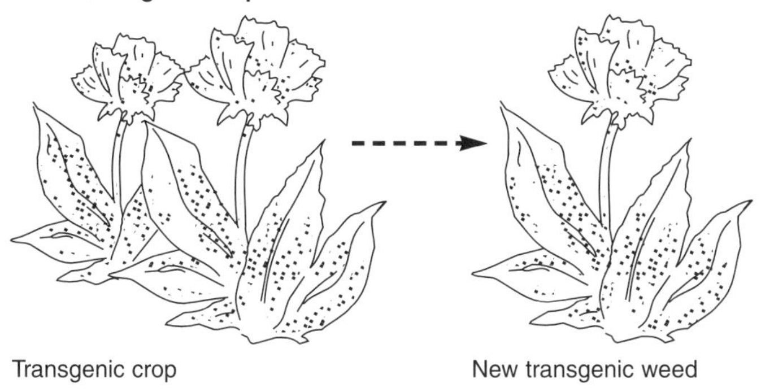

2. The transgenic crops transfer pollen to wild relatives that become weeds.

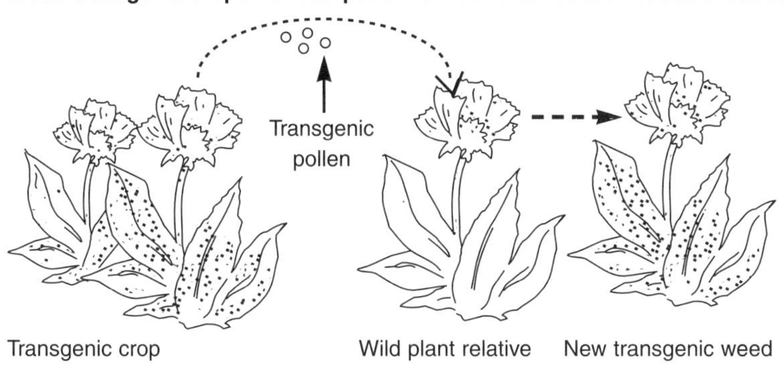

Figure 2.1
Risks of transgenic crops

may not have undesirable impacts. But in rare circumstances, the modification may enable the crop to become a weed that is expensive to control and impossible to eradicate.

The story of kudzu's introduction into the United States is instructive about the damage that can ensue when a nonindigenous plant finds a successful ecological niche in a new environment (Alabama Cooperative Extension Service, no date). Kudzu was first used in the United States in the late 1800s as an ornamental vine to shade the porches and courtyards of southern homes. Promoted in the early 1900s as a forage crop and widely planted to reduce soil erosion in the 1930s, kudzu soon spread out of control. Today, kudzu infests an estimated 7 million acres in the southeastern United States, despite repeated attempts to remove it.

The second category of risks concerns the transfer of transgenes to crop relatives. It is important to note that novel genes added to crops will not necessarily stay put. Where crops are grown in proximity to wild relatives, the novel genes can be transferred into wild plants. An example would be the transfer of a gene for drought tolerance to a wild relative of sunflower. That newly drought-tolerant relative might then be able to invade an arid habitat, displacing native plants. If the displaced plants offered a better habitat for small mammals than the invaders, the harmful effect could ripple beyond the arid habitat.

A third category might be considered a subset of gene transfer, but because it involves independently reproducing entities it seems to deserve special treatment. The risks concern the instances where the transgene added to a crop plant is a component of a virus. In such cases, there is a possibility of creating new viruses that might intensify or cause new plant diseases.

It is difficult to assess which of these environmental risks of transgenic crops are of greatest concern. Not only are scientists and farmers without much real experience with such organisms, but we are envisioning risks that result from hard-to-predict conjunctions of organisms and often changing environments. In rough terms, of the known risks, the risks from gene transfer are likely to be of the greatest concern and the production of new viruses the least. But no rigorous comparison has been done, and it would not be surprising if experience provided midcourse corrections to this view.

Field Tests Tell Little about Commercial-Scale Risk

Some may assume that the hundreds of field trials of genetically engineered crops conducted so far constitute a record of safety that should allow the industry to proceed confidently to commercial use. But the safety record of small-scale field tests, as welcome as it is, does not necessarily predict safety on a commercial scale. As discussed above, the ecological risks of transgenic crops depend on relatively rare events occasioned by the interaction of particular engineered plants with a particular environment. In general, the lack of such events in field tests does not predict safety on a larger scale, because the opportunities for such interactions are severely restricted. Commercial-scale use vastly increases the opportunities for the rare harmful conjunctions of factors to occur. This is true in three respects.

First, commercialization significantly increases the potential for exposure to transgenic crops. The number of plants involved in commercial use is orders of magnitude larger than the number involved in field tests. Most field tests involve releases on no more than ten acres. By contrast, commercial use of major crops like corn and soybeans could entail cultivation on millions of acres in the United States alone. Second, field tests are conducted under conditions that severely limit the escape of plants or genes from the test plots. For example, as a rule, test sites are far removed from other crops or wild relatives with which the transgenic crop can interbreed. Also, sites are monitored to detect transgenic crops that escape. No such restrictions will apply to commercial use, however. Once available on the market, engineered plants will be free to migrate away from the farm, and their pollen will flow unimpeded to relatives in agricultural and nonagricultural habitats. Third, in comparison to field tests, commercial use will involve crops planted in far more diverse environs, in proximity to a broader selection of relatives, in different climates, and subject to a greater variety of weather events, such as floods and hurricanes. Events like floods, for example, can expose seeds and plants to many new, potentially congenial, environments.

Risks Are Global

Agricultural biotechnology is a global industry. The international nature of the seed trade means that crops approved for sale here are likely to be sold and grown in other countries. International distribution exacerbates the environmental risks by increasing the variety of environments and conditions in which crops will be grown.

It also adds a new dimension to the magnitude of environmental risk by adding to the list of affected populations the many unique centers of genetic diversity for agricultural crops that are found outside the United States. These unique plant populations are vital to U.S. and world agriculture, and the United States must consider its responsibility to protect these centers, which are already under considerable threat.

U.S. responsibility for impacts outside this country may be particularly great if U.S. government approval for the sale of transgenic plants is regarded as a general seal of safety. Such an occurrence would be scientifically misleading. Any determination that a transgenic crop is safe to grow in the United States depends on U.S.-specific conditions, particularly the distribution of wild relatives, and would carry no implication of safety for use elsewhere in the world. If the United States believes that its reviews are, in effect, the only ones crops will receive before they move into international commerce, it must take greater steps than it might normally have done to protect the global treasures represented by centers of diversity.

3

Environmental Risks Posed by Transgenic Crops

The commercialization of transgenic crops may pose a spectrum of risks—from ill effects on humans and animals that consume engineered crops to the disruption of wild ecosystems. This chapter focuses exclusively on two major categories of environmental risks: those of the engineered plants themselves and those associated with the movement of transgenes into other plants.

First, engineered plants risk becoming weeds in agricultural ecosystems or becoming established outside the field, disturbing unmanaged ecosystems.[1] Weediness, as used here, covers any undesirable effect of plants, including the initial impact of a new weed, other effects such as reduced biodiversity, nontarget effects of pesticidal and pharmaceutical transgenes, secondary effects on agricultural ecosystems, as well as other cumulative and cascading impacts. The second category of risks concerns the movement of new genes to crop relatives. Where crops are grown in proximity to wild relatives, novel genes can be transferred into the wild plants. Transgenic wild plants may then be capable of invading farmers' fields or altering natural ecosystems. When the transgene added to a crop is a component of a virus, there is a possibility of creating new viruses or altering the host range of existing viruses.

This chapter first addresses the risks of the transgenic crops themselves, and then details the implications of transgene flow from the transgenic crop to wild/weedy relatives.[2] Finally, it discusses risks posed by transgenic virus-resistant crops, exploring the possibility that some risks have yet to be identified.

Potential Adverse Impacts of Transgenic Crops

Transgenic Crops May Become Weeds

One of the chief concerns with engineered crops is that they may become weeds in agricultural and nonagricultural settings. This section describes the concept of weeds and their characteristics, discussing the potential for transgenic crops to become weeds in farmers' fields or to invade nonagricultural habitats.

Definition of Weeds. "Weeds," in common usage, are plants that happen to be in the wrong place at the wrong time. Indeed, no plant is intrinsically a weed—the designation depends on both context and human values. Thus, the same plant may be a weed in one situation and a desirable plant (such as a crop or lawn grass) in another. Not surprisingly, there are a number of definitions of weeds. In a widely used definition, Baker (1965) focuses on where weeds are found: "[A] plant is a 'weed' if, in any specified geographical area, its populations grow entirely or predominantly in situations markedly disturbed by [humans] (without, of course, being deliberately cultivated plants)." A broader, more human-oriented, definition considers a weed to be "a plant not intentionally sown, whose undesirable qualities outweigh its good points" (Granatstein, 1988). The Weed Science Society of America (WSSA) (1989) similarly defines a weed as "any plant that is objectionable or interferes with the activities or welfare of [humans]." In this discussion, the term "weed" is used more like Granatstein and the WSSA—to broadly cover unwanted plants. It includes plants that reduce a farmer's yield as well as those that invade nonagricultural habitats such as home lawns, national parks, waterways, and wildlife habitats. The term "wild" is reserved for plants that grow exclusively or predominantly in habitats not heavily influenced by humans.

Origin of Weeds. Plants that are considered weeds may be species native to the region where they are found or they may have been introduced from other habitats. Some may be crops, like Bermuda grass, which are cultivated in some areas but are considered weeds in others. Other weeds may be relatives of crops. Nearly all food and fiber crops, such as rice, barley, sorghum, and some tree crops, have close relatives that are regarded as weeds somewhere in the world (Harlan, 1982).

Harmful Exotics in the United States

Harmful exotic organisms, many brought into the United States accidentally, are emerging as a major environmental problem. Newcomers like kudzu, purple loosestrife, and gypsy moth have wreaked havoc in native ecosystems, caused disease and natural disasters, and cost billions of dollars in damage and control costs.

Over the last decade, scientists and others have debated whether genetically engineered organisms are more comparable to exotics or the routine introduction of new varieties of crops or breeds of domestic animals, which are also genetically novel. To put a complicated debate simply, considering engineered organisms to resemble agricultural breeding leads to predictions of greater environmental safety than a comparison to exotic organisms.

A report from the U.S. Congress Office of Technology Assessment (OTA), *Harmful Non-Indigenous Species in the United States* (U.S. Congress OTA, 1993), offers a new framework for this debate. Rather than contrasting agricultural introductions and the invasion of exotic organisms, the OTA encompasses both under a more comprehensive category, nonindigenous species.

According to the report, nonindigenous species are organisms living beyond the geographic area that they would inhabit if they were not affected by significant human influence. The broad designation covers all domesticated organisms and encompasses much of agriculture. In addition, the definition includes feral plants and animals, organisms whose ancestors were once domesticated, but are now living free (an example would be Johnson grass, a tenacious weed once grown for forage). It also includes all hybrids except those formed naturally between indigenous species, for example, corn and wheat hybrids. Under the OTA scheme, *all* genetically engineered organisms are included in the broad category of nonindigenous species.

The OTA report acknowledges that most nonindigenous species are beneficial, and in fact are the backbone of world agriculture. But the report also emphasizes the enormous damage done by the many that are harmful. Some, like the imported fire ant, threaten human health. Others, like the boll weevil, cause millions of dollars in crop losses every year. Still others, like purple loosestrife, disrupt ecosystems and displace native organisms. Many of the harmful nonindigenous species, for example kudzu, were originally introduced purposefully. Although they represent a relatively small portion of the total, harmful nonindigenous organisms affect all regions of the country, and do substantial damage. Control costs, where control is an option, run into the billions of dollars.

The OTA report recasts rather than resolves the debate between the agricultural and the exotic models. It puts agriculture into the proper context as a source of both beneficial and harmful nonnative species, gives new urgency to the protection of indigenous flora and fauna, and makes it clear that genetically engineered organisms are among the many kinds of nonindigenous species that pose environmental threats. The continuous stream of new nonindigenous organisms carried in ballast water, backpacks, and airplane cargo constitutes a major environmental threat to ecosystems in the United States. According to our current understanding, the problems caused by the intentional introductions of genetically engineered organisms in agricultural settings should enlarge that threat by only a small amount. On the other hand, agricultural introductions—whether of engineered or traditionally bred organisms—may constitute the part of the nonindigenous organism problem most amenable to control.

Many plants purposefully introduced as food and forage crops or ornamentals have later become weeds.[3] Table 3.1 lists a number of introduced plants that have become serious weeds in the United States. Crabgrass and Johnson grass are two notable examples. Large crabgrass (*Digitaria sanguinalis*) was introduced into the United States twice—around 1850 and 1900—to produce grain.[4] Even though it was soon replaced by corn and wheat as a crop, crabgrass persisted as a serious weed in and around fields. It also invaded other habitats (Foy et al., 1983 and references therein). Johnson grass (*Sorghum halepense*), brought into the United States in the early nineteenth century as a perennial forage grass, is now considered one of the world's ten worst weeds (Foy et al., 1983 and references therein).

Importance of Weeds. The fact is, the term "weeds" covers plants unwanted in farms, lawns, roadsides, and unmanaged ecosystems that have become a multi-billion-dollar problem. In 1991 alone, farmers and others in the United States spent over $4 billion to control weeds (Aspelin et al., 1992). Serious economic pests, weeds also cost growers around the world untold millions of dollars annually in decreased yield and seed quality. Weeds reduce the growth and yield of crops and other plants. They compete for water, light, and nutrients; produce toxic com-

Table 3.1
Examples of intentionally introduced plants that have become serious weeds in the United States

Plant	Purpose of introduction	Areas invaded
Artichoke thistle	Food crop	Rangelands
Bermuda grass	Forage, turf	Pastures, orchards
Cogongrass	Forage	Southern farms
Crabgrass	Food crop	Pastures, croplands
Dalmatian toadflax	Ornamental	Rangelands
Dyers woad	Dyes	Rangelands, croplands
French tamarisk	Ornamental	Pastures, flood plains, waterways
Hydrilla	Aquarium trade	Lakes, reservoirs, waterways
Japanese honeysuckle	Ornamental	Pastures, woodlands
Japanese knotweed	Ornamental	Lowlands, homesites
Johnson grass	Forage	Croplands
Kochia	Ornamental	Widespread
Kudzu	Ornamental, forage, erosion control	Forests, rights of way, field borders
Melaleuca	Ornamental	Wetlands
Multiflora rose	Erosion control, fencerows	Pastures
Purple loosestrife	Ornamental	Lakeshores, wetlands
Reed canarygrass	Forage	Canal and ditchbanks
Strangler vine	Ornamental	Citrus
Water fern	Ornamental	Waterways
Water hyacinth	Ornamental	Lakes, waterways
Wild melon	For observation	Imperial Valley, croplands
Yellow toadflax	Ornamental	Rangelands

Sources: Modified after Williams, 1980; Turner, 1988; P. Regal, personal communication, 1992.

pounds; even harbor insect pests or plant pathogens. Where crops are grown for seed (for example, wheat, barley, oats, rice), weed seeds, harvested along with the crop, reduce crop seed quality. Weeds may also decrease the grazing quality of pastures. A number of annual species of oats, barley, and bromegrass, for example, have become major weeds in some California grasslands, successfully reducing the quality of forage for livestock (Turner, 1988 and references therein). Weeds interfere with transportation in waterways and along roads and rights-of-way. Water hyacinth (*Eichhornia crassipes*), introduced into the southeast United States from South America for ornamental purposes, has now invaded waterways from Virginia to California (Foy et al., 1983 and references therein; Williams, 1980). Waterways in Florida became clogged, severely impeding navigation, just twenty years after the introduction of the water hyacinth. The state of Florida spends $7 million annually combatting hydrilla (*Hydrilla verticillata*), a plant introduced from South America in the late 1950s for use in aquaria. Hydrilla now chokes a number of lakes in the state (U.S. Congress OTA, 1993 and references therein; Williams, 1980 and references therein).

Weeds may also have adverse health impacts on humans and other animals. Some are poisonous if consumed or cause dermatological or respiratory allergenic effects. The Brazilian peppertree (*Schinus terebinthifolius*), for example, was introduced into the United States from Brazil for its showy foliage. The tree, however, has proven an aggressive invader. Moreover, its pollen is a respiratory irritant to some people, its sap causes skin blisters and rashes, and the poisonous fruit has killed raccoons, deer, and horses—and caused illness in children (U.S. Congress OTA, 1993 and references therein; Williams, 1980 and references therein). Halogeton (*Halogeton glomeratus*), a plant toxic to sheep, was inadvertently introduced into the western United States over sixty years ago. Within twenty-five years of its discovery in Nevada in 1935, it had spread over millions of acres of rangeland in five states. Thousands of sheep have died from the toxic oxalates produced by the plants (Williams, 1980). Other weeds may have more indirect effects. For example, some aquatic weeds enhance mosquito-breeding habitats, allowing larger populations of pathogen-carrying insects to develop.

To decrease the impacts of weeds, U.S. farmers, landscapers, home gardeners, foresters, resource managers, and government agencies spend billions of dollars every year. For example, the Environmental Protection Agency estimated that 628 million pounds of herbicides (active ingredients) were purchased in the United States at a cost of over $4.3 billion in 1991 (Aspelin et al., 1992).

Weeds are also problems in settings less dominated by humans. They may invade local ecosystems, displacing native plants, and perhaps permanently altering the makeup of the community. For example, purple loosestrife (*Lythrum salicaria)* was introduced from Europe in the last century and is now replacing native wetlands vegetation.

Weediness Traits. Because the term is a contextual one, no plant can be said to be a weed by nature. But weeds often appear to have particular physiological and structural traits that allow them to persist in environments managed or otherwise influenced by humans. Some of these traits also allow them to compete successfully against cultivated crops or other plants. Different weeds exhibit different combinations of traits to confer their competitive advantage. Some plants may be weeds primarily because of their vigorous root growth and high seed production. Others may be notable for their capacity to reproduce vegetatively, choking out other plants. (Table 3.2 lists some characteristics often associated with weedy plants.)

Major Weed Strategies: Persistence and Invasiveness. A plant species may become a weed through one of two general strategies: it may persist

Table 3.2
Characteristics often associated with weediness

1. Seeds germinate in many environments.
2. Seeds remain viable a long time.
3. Plants grow rapidly through vegetative phase to flowering.
4. Plants produce seeds continuously as long as growing conditions permit.
5. Flowers are self-compatible, but not obligatorily self-pollinated.
6. Pollen from flowers that are cross pollinated is carried by nonspecialized flower visitors (usually insects) or by wind.
7. Plants produce large numbers of seeds in favorable environmental circumstances.
8. Plants produce seed in a wide range of environmental circumstances.
9. Plants are adapted for both long-distance and short-distance dispersal.
10. If perennials, the plants have vigorous vegetative reproduction or regeneration from fragments.
11. If perennials, the plants are brittle near the soil line to prevent easy withdrawal from the soil.
12. Plants compete by special means, such as forming rosettes, choking growth, producing toxic chemicals.

Source: Modified after Baker, 1965, 1974.

and be a problem where it is introduced, or it may invade and alter other habitats. The first strategy is probably best illustrated by the case of a crop becoming a weed in a subsequent crop in the same field. For example, Bermuda grass planted as a forage may persist in subsequent plantings of other crops. Volunteer potatoes are weeds the next season in fields in the northwest United States (Smid and Hiller, 1981; Thomas and Smith, 1983).

The term "invasiveness" describes a plant's capacity to spread beyond the site of introduction and become established in new settings. Spread may occur via seeds or vegetative parts (for example, roots, rhizomes, stolons). Melaleuca (*Melaleuca quinquenervia*), or paper bark tree, was introduced from Australia nearly a century ago to dry out part of Florida's Everglades swamps (Ewel, 1986; U.S. Congress OTA, 1993). An aggressive invader, melaleuca has swept across south Florida in the last thirty years, infesting 450,000 acres. It is replacing cypress forests, sawgrass marshes, and other natural habitats at the rate of fifty acres a day. In addition to tolerating a broad range of conditions, including flooding, drought, salinity, and fire, melaleuca reproduces prolifically both vegetatively and by seed. It is considered to be Florida's most disastrous deliberate introduction of the twentieth century (U.S. Congress OTA, 1993, and references therein).

The distinction between persistence and invasiveness is relevant in the evaluation of potential impacts of transgenic crops. In essence, the distinction tells where to look for effects and what traits to look for, particularly the capacity for flexibility or variation in certain attributes. To evaluate persistence, the focus is on subsequent growing seasons in the same field where the transgenic crop was planted. To assess invasiveness, the transgenic plant's capacity to disperse and establish in adjacent and nearby habitats is assessed.

Transgenic Crops as Weeds. Scientists have suggested that some transgenes may confer or enhance weediness in some crops—that is, they may enhance the crop's capacity to persist in a field, invade new habitats, or both (Crawley, 1990a; Williamson, 1992). As shown above, weeds have serious economic and ecological consequences. The possibility that engineering will convert crops into new weeds is a major risk of genetic engineering.

The kinds of risks associated with transgenic crop weediness are similar to those presented by the introduction of nonindigenous plant species into an environment. Most nonindigenous plants do not survive in new environments, but the relatively few that do survive can cause significant ecological disruption. A U.S. Congress Office of Technology Assessment report concludes that nonnative plants have caused hundreds of millions of dollars of economic losses in the United States alone (U.S. Congress OTA, 1993). Most genetically engineered plants would not be expected to become weeds; those that do, however, could present serious problems. Some scientists use the so-called ten-ten rule to describe the success rate of invading plant species (Williamson, 1993). According to the rule, roughly 10 percent of the species found as free-living individuals are likely to establish self-sustaining populations and 10 percent of those are likely to become pests. A recent study of crops in Britain indicated that agricultural crops were much more likely than the typical imported plant to escape and survive as the free-living individuals from which the ten-ten calculation begins (Williamson, 1994).

Weediness can be enhanced by alterations of numerous traits. Persistence can be affected by altering traits such as seed dormancy, seed germination, tolerance to biotic (living organisms) or abiotic (nonliving) stresses, or competitiveness of vegetative plant parts. The timing of a crop's appearance in the next growing season might also affect its weed status. A crop in which transgenes affected seed germination—conferring rapid germination in cooler spring temperatures, for instance—might become a successful weed competing against another crop planted subsequently in the same field, but germinating more slowly. Transgenes could affect such traits either directly or indirectly.

The addition of transgenes may also enhance a crop's ability to invade other habitats, such as fields, meadows, or forests adjacent to cultivated sites. Models of invading organisms show that the rate at which a weed moves into a new habitat is directly proportional to its dispersal and rate of population growth (Andow et al., 1990). Thus, any modification that enhances population growth, such as increased reproductive capacity or survival, theoretically increases invasiveness.

Genetic engineering might alter invasiveness by affecting properties such as seed germination and dispersal, seedling growth, or root growth.

These different conditions may include more or different competitors and pests or tolerance to a broader range of abiotic factors—temperature, moisture, salt. For example, if seeds of a transgenic crop made tolerant to cold temperatures were dispersed to colder climates, the crop might eventually invade habitats heretofore not colonized by the plant.

Could Transgenes Convert Crops to Weeds? How likely is it that adding one or a few transgenes will make crops weedy? A widely cited notion holds that weediness is a complex trait and that, in general, the addition of one or a few genes is unlikely to convert crops that are nonweedy to weediness (Crawley, 1990a; Hauptli et al., 1985; Keeler and Turner, 1991; National Research Council, 1989b). Since the current limit on the transferred genetic material is about two to five genes, this notion has provided considerable reassurance that weediness risks from transgenic crops are small. Although the argument has some force, it goes too far if offered as a blanket reassurance that genetic engineering restricted to several genes cannot convert a crop to a weed.

The argument depends on two assumptions: (i) that a specific set of traits is necessary to make a plant a weed and (ii) that most crops are so nonweedy that they lack most of the required traits. Much of the idea that weeds exhibit a specific set of traits derives from a list of the characteristics of an ideal weed published in 1965 by a weed scientist, H. G. Baker. For some, Baker's list of characters has become a kind of quantitative index of weediness: the more traits on the list a plant possesses, the worse the weed. Conversely, if a nonweedy plant has few of Baker's characters, many new traits would be needed for it to become a weed. An industry analysis of the potential weediness of transgenic virus-tolerant squash, for example, argued that since squash possesses only three of Baker's characters, addition of the virus-tolerance gene would be unlikely to convert squash into a weed (Asgrow Seed Company, 1992).

Use of Baker's list to predict weediness has been strongly criticized in the scientific literature. In a study of forty-nine British annual plants, Williamson (1993) showed that although Baker's characters are weakly correlated with weed status, they have little predictive force. The study found that serious weeds can have few Baker characters, while plants with many characters are weak weeds (Williamson, 1993). Moreover, the most seriously weedy plants in the study have only three, four, or five of nine[5] of Baker's characters (Perrins et al., 1992; Williamson, 1993). Having rejected Baker's list, Williamson (1994) tried to find other characteristics predictive of weediness by analyzing fourteen major British plant pests. He concluded that there are no characters essential to weediness. "Successful pests may have any life-form, can occur in any habitat, may have been introduced for all sorts of reasons, may come from anywhere and may have any form of reproduction" (Williamson, 1994).

The notion that crops cannot be converted to weeds also assumes that crops are so nonweedy that they possess few, if any, of the set of weediness traits. This assumption is only partially true. Many crops, in fact, are weeds. For such crops, the issue is whether the addition of a few genes could intensify weediness, not confer it. Of course, some crops appear to be extremely nonweedy; some in fact are so ecologically debilitated that they do not survive without human assistance. Corn, for example, has never become established as a wild plant in the United States despite hundreds of years of planting. Millennia of breeding have transformed corn into a crop that is dependent on human intervention for survival and productivity. It is difficult to imagine a one- or two-gene transfer that would enable corn to displace other plants in unmanaged settings.

But few crops are as ecologically debilitated as corn. Some nonweedy crops might possess a nearly complete complement of whatever traits the crops would need to be a weed. Where such crops are on the edge of weediness, the addition of one or a few adaptive genes could push them over it.

The challenge is to determine which nonweedy crops fall into "almost a weed" category. One reasonable approach to this issue is to look at the set of crops that are similar or closely related to known weeds. Many crops, such as alfalfa and other forages, are similar in genetic makeup to wild plants (Regal, 1990). This fact may mean that their shared genes extend to those for weediness traits. A number of other crops, including barley, lettuce, oats, potatoes, sorghum, and wheat, have close weedy relatives (Harlan, 1982). For example, red rice, a close relative of the cultivated crop, is a major pest of rice fields. The cucumber/squash family (*Cucurbitaceae*) contains many crop, weedy, and wild members. The Brassica group—broccoli, cauliflower, cabbage, radish, mustard—includes several weeds, some of which are closely related to the crops.

Still other crops are actually weeds in certain situations. They already possess traits needed to confer weediness. Bermuda grass, an important turfgrass and forage crop, is also a serious weed in some areas of the country (Ellstrand and Hoffman, 1990). Sunflowers, which are native to the United States, are troublesome weeds across the midwestern part of the country. Strawberries have become a weed problem in some perennial and no-till systems in the U.S. Pacific Northwest. Annual ryegrass is a major weed in winter wheat in areas where the ryegrass is grown as a crop; persistent broccoli is a problem in broccoli-growing areas of the United States (M. L. Roush, personal communication, 1992). Thus, many crops are sufficiently closely related to weeds that it is reasonable to assume that some small changes might move them into that category.

Pleiotropic and Epistatic Effects of Transgenes

Genetic engineers generally have a specific goal in mind when they splice transgenes into plants. They add insect-toxin genes to enable a plant to repel insects, herbicide-tolerance genes to allow a crop to grow despite applications of herbicides, or oil modification genes to obtain a new profile of oils in seeds. However, it is well known that in addition to the expected effects of a transgene, a new gene may also alter the characteristics of a plant in other, less predictable ways—through pleiotropy and epistasis.

Pleiotropy means that a single gene product can affect more than one trait. For example, some corn cultivars carry a natural corn gene that interferes with the production of pollen grains, causing male sterility. That same gene makes the corn lines susceptible to a fungal pathogen that causes southern corn leaf blight. In 1969 and 1970, when more than three-fourths of the U.S. corn acreage was planted with male-sterile lines, southern corn leaf blight caused severe losses in the South and the Corn Belt (Levings, 1990 and references therein). Through traditional methods, breeders altered the corn genome to avoid a repetition of the epidemic.

Epistasis refers to the capacity for one gene to modify the expression of another gene that is not an allele of the first. (Alleles are different versions of a gene for a particular trait.) An example is the production of purple flowers when two pure-bred, white-flowered sweetpea varieties are crossed. The purple color results from the interaction of two separate genes for color—neither one of which can produce the purple color alone (Curtis, 1983).

It is reasonable to expect that some pleiotropic and epistatic effects will also occur in some transgenic crops. The difficulty of predicting these effects compounds the complexity of analyzing risks of engineered plants.

Thus, crops are not necessarily separated from weediness by large numbers of traits, and many crops are already weeds or close to being weeds. The theory of the complexity of weediness does not justify complacency about one to two gene additions to crops. Concern that one or two genes could produce ecologically significant traits that might qualify plants as weeds is heightened by direct observations of situations in which one or a few genes have altered traits that contribute to a population's ecological success. For example, a single or a few disease- or insect-resistance genes can make the difference between a plant's surviving a pest or being decimated by it, as in late blight disease of potatoes and stem rust disease and Hessian fly in wheat (Day, 1974). Plant breeders may incorporate a single gene or a few genes, if they can find them in the crop or its wild relatives, to obtain resistant crop cultivars. Plants without the resistance gene(s) and in the presence of the pest or pathogen may not survive or would yield less if they did survive.

Jain and Martins (1979) found that a single gene change in rose clover (*Trifolium hirtum*) enhanced the plant's invasiveness. The clover's success in colonizing roadsides was associated with a larger and hairier calyx[6] which, unlike noncolonizers, remained attached to germinating seedlings. The attached calyx apparently accounted for higher germination on the roadsides.

The presence of a single suppressor gene in cultivated oats (*Avena sativa*) distinguishes the crop from weedy, wild oats (*A. fatua*). In the crop, the suppressor gene prevents expression of "fatuoid" characters, the ability to retain seed at maturity and the presence of awns (bristle-like structures) and pubescence (downy, fine hairs) at the base of the grain (Thomas and Mytton, 1970).

Small differences in resistance to fungal infection in skeleton weed (*Chondrilla juncea*), a weed introduced in cereal crops in Australia, dramatically affected the distribution of the three forms of the pest. Before 1970, form A predominated in southeast Australia. With the introduction of a rust fungus to control the weed in the early 1970s, form A declined and forms B and C became more widely distributed. In this case, form A was susceptible to the biological control agent and forms B and C were not (Burdon et al., 1981). These small changes are likely to be controlled by one or few genes.

In sum, the suggestion that the transfers of one or two genes to crops are unlikely to create weeds is not applicable to all crops. For crops that are already weeds or closely related to weeds, the addition of one or a few transgenes conferring a significant ecological advantage might well be enough to push it over the edge to increased persistence and/or invasiveness. In the United States, plants that may fit this description include alfalfa, some ornamental bamboo, Bermuda grass, blackberry, radish, rapeseed, raspberry, and sunflower. Moreover, genetic engineers will not be confined to one or two gene transfers for long. Eventually they will add many different transgenes to crops, which will increase the odds that some transgenic crops will become weeds.

Other Aspects of Weediness of Transgenic Crops

As noted earlier, weediness is a broad term that covers unwanted plants in a variety of settings. This section focuses on some aspects of weediness that occur at times or places well removed from the initial perturbation in the ecosystem. Such delayed effects include impacts that cascade through or accumulate in ecosystems, effects on nontarget organisms of pesticide- and pharmaceutical-producing transgenics, and secondary impacts on farm management. These different aspects of weediness should not be seen as mutually exclusive or rigid categories, but rather as a continuum of effects that transgenic crops may produce.

Initial Impacts. A transgenic crop that becomes a weed may have one or many of a broad range of impacts. The initial effect, which may subsequently reverberate to cause other problems, may well be the crop's capacity to persist, unwanted, in a field or a pasture or to invade a wild habitat near a farm. Once a transgenic crop becomes established, other aspects of weediness—nontarget effects, altered community makeup, impacts on farm management—may come into play over time.

Cumulative and Cascading Effects. Some effects of genetic engineering discussed earlier may accumulate and/or cascade in ecosystems. One early impact may set off a series of steps that sequentially amplify the magnitude of the impacts. The invasion of a transgenic crop into a new habitat or the effect of a transgene product on a nontarget organism may initiate a series of events—as some of the examples in the previous sections illustrate. Repercussions may show up, not after the first or second transgene or transgenic plant, but from the impacts of several or many acting alone or in concert.

A well-known example of cascading and cumulative effects in nature is the magnification of the insecticide dichloro-diphenyl-trichloroethane (DDT) as it passes through the food chain from aquatic invertebrates to fish to avian species (Graham, 1970). DDT, by reducing the reproductive success of many birds, causes reverberations through ecosystems. Similarly, transgenic crops may initiate a series of events that ripple throughout an ecosystem. For example, transgenic crops that are able to invade nonagricultural habitats may be competitively superior to other members of the community and may replace native species—with the result that the makeup of the community is altered or local species or subpopulations are driven to extinction.

A simple hypothetical example of genetic engineering's potential to ultimately alter community structure might begin with transgenic salt-tolerant rice planted near coastal wetlands. It is conceivable that the rice could invade the salt-water ecosystems, displacing native salt-tolerant species. As the native populations decline, other organisms typically associated with them—such as algae, microorganisms, insects, other arthropods, amphibians, birds—might not be compatible with the invading rice. Different organisms, new to the salt-water marsh, might find homes in the new rice-dominated ecosystem.

These kinds of complex adverse impacts have received little attention from genetic engineers or regulators. Perhaps the neglect is due to the difficulty of predicting the effects and the belief that ecosystems can be manipulated with impunity. In fact, the putative resilience of ecosystems may have lulled humans into complacency because many perturbations are absorbed and "defused" before they have a chance to cascade into disaster. It is believed that, through feedback, natural systems are capable, to some extent, of "pulling themselves up by their own bootstraps" (Perry et al., 1989).

Yet there are limits to this natural resilience. Species with transgenic traits, particularly those that make major alterations in function, incorporate multiple traits, and add traits from dissimilar organisms, may well stress natural systems, perhaps beyond their ability to react and recover (M. L. Roush, personal communication, 1992). If gene-transfer technology succeeds as its supporters predict, global agriculture and the world's ecosystems will be confronted with hundreds of transgenes in the major food and fiber crops and in many minor crops. The potential cumulative effects are simply not known.

Nontarget Effects of Transgenic Pesticidal and Pharmaceutical Products. Some kinds of transgenic crops raise special issues because of the potential ecotoxicity of their novel gene products. Where plants are engineered to manufacture pesticides and drugs, it is to be expected that the plants and their novel products will harm organisms other than their intended targets.

For example, pesticides are rarely selective enough to kill only the target pest and as a rule pose risks to nontarget organisms as well. Transgenes for insecticidal or fungicidal compounds introduced into crop plants to inhibit pests, may kill nontarget, and even beneficial, insects and fungi. Soil-inhabiting insects that degrade plant debris containing the insecticidal toxins produced by the soil bacterium *Bacillus thuringiensis* (Bt) may be harmed by the toxin[7] (Palm et al., 1994). This is particularly important because genetic engineers are currently adding genes for Bt toxins to a wide array of crops to enable the plants to produce their own insecticide (Union of Concerned Scientists, 1994).

In addition, fungicidal transgene products such as the enzyme chitinase may reduce populations of mycorrhizae, fungi that extract nutrients from the soil for the benefit of plant roots with which they are intimately associated. The cell walls of the fungi contain chitin, which is broken down by chitinase. Once the walls are degraded, the fungal cells die. Loss of fungi involved in resource cycling could impede the flow of nutrients vital to ecosystem functioning.

Similarly, where transgenic crops are used to manufacture human and animal drugs, hormones, and vaccines as well as industrial enzymes, oils, and other chemicals, it is reasonable to expect that some will be harmful to other organisms. In fact, little is known of the potential nontarget effects of these kinds of plants. For example, next to nothing is known of the effects of "immunofoods"—crops engineered to produce edible vaccines for a particular human or animal pathogen—on soil organisms or animals that feed on the crops. Wild animals may consume alfalfa producing anticancer drugs, growth hormones, and vaccines—with unknown consequences. The effects on soil microorganisms, insects, and earthworms, exposed to the chemicals as crop residue decays, are likewise not known.

Secondary Effects in Agricultural Ecosystems. In addition to the possibility that some engineered crops may persist in their fields and compete with other crops, farmers may have to deal with less direct impacts that require changes in their farm management practices. For example, transgenic Bt-containing crops may have an impact on the efficacy of Bt, a relatively safe biological insecticide. There is general concern (U.S. Department of Agriculture, 1992) that widespread use of Bt-containing crops could accelerate the development of insect pest resistance to Bt. Already, eight species of insects have developed resistance to Bt toxins, either in the field or laboratory, including diamondback moth, Indianmeal moth, tobacco budworm, Colorado potato beetle, and two species of mosquito (McGaughey and Whalon, 1992 and references therein). The problem with transgenic Bt plants is that they will increase the exposure of pests to Bt, especially where the Bt toxin is continuously expressed in a plant throughout its growing season. This is significant because long-term exposure to Bt toxins promotes development of resistance in insect populations. This kind of exposure could lead to selection for resistance in all life stages of the insect pest on all parts of the plant for the entire season (McGaughey and Whalon, 1992).

Unless resistance management strategies are developed and implemented, growers may soon lose the benefit—reduced synthetic insecticide use—that Bt-crops and sprays bring. With the loss of Bt's efficacy, farmers—particularly organic farmers who depend on Bt—will be faced with decisions on how to reduce resurgent damage by insect pests. Thus far, even though research on resistance management strategies has yet to yield detailed, workable strategies, transgenic Bt-crops have been approved for commercial production.

The problem of evolution of pest resistance is not confined to Bt and its target insects. Other pests—fungi, bacteria, viruses, insects—may also evolve resistance to new pest-control genes engineered into plants.

Summary

Transgenic crops pose risks of becoming weeds within fields or in nearby habitats. In addition to these initial impacts, transgenic crops present risks that may be felt at a place or time distant from initial perturbations. These include cascading and cumulative effects, effects of the transgene product on nontarget organisms, and secondary impacts in agricultural ecosystems. With current understanding and methods, the initial impact of transgenic crops—as expressed through their ability to outcompete nontransgenic forms—is the only assessable aspect of weediness. An assessment scheme to predict the capacity of transgenics to invade ecosystems is described in the next chapter. For now, all other aspects of weediness remain difficult to predict and evaluate.

Potential Adverse Impacts of Transgene Flow to Other Plants

In addition to the risk that transgenic crops might become weeds, large-scale releases of transgenic crops risk transferring transgenes from crops to other plants, which may then become weeds. This transfer of genetic material is called "gene flow."

Among biologists, gene flow into a plant population may refer to the introduction of genes by pollen, seed, or asexual propagules. This section considers only the flow of genes via pollen during sexual reproduction; thus, the term "gene flow" here means only the flow of genetic material from one population to another via pollen.[8] Wild/weedy plants that are recipients of transgenes by gene flow are referred to as "transgenic wild/weedy" plants.

This section addresses the likelihood that gene flow will occur, the factors affecting it, and its potential adverse repercussions. The weediness risks presented by transgenic wild/weedy plants are generally the same types as those identified for the transgenic crop. The level of risk, however, could be higher for the wild relatives because they are not ecologically debilitated as are some crops. Finally, this section briefly considers the largely unknown field of nonsexual, or horizontal, gene flow.

Likelihood of Occurrence

Once transgenic crops are planted in large numbers near sexually compatible wild/weedy relatives or other crops, transgenes will almost certainly flow via pollen to these other plants (see "Sexual Reproduction in Flowering Plants" and figure 3.1). Seeds of wild/weedy plants and transgenic crop hybrids[9] will form and many will contain transgenes. Generally, half the hybrid's genes will come from the transgenic crop and half from the compatible crop or wild/weedy parent.[10]

If these hybrids produce viable, fertile seeds that grow into plants capable of sexually reproducing with the weeds that surround them, crop genes—including the transgene—may be introduced into subsequent generations of the wild/weedy population. This process of introducing new genes into a wild population via hybridization and backcrossing is called introgression. The first step in introgression is the formation of the wild/weedy plant x crop hybrid. Then, in the second phase of the process, that hybrid may cross with the wild/weedy parent (a backcross), moving the gene into the wild/weedy population.

The result is the introduction of crop genes into the gene pool of the wild/weedy plant population. After the first backcross most (75 percent on average) of the genes in the progeny plants will have been contributed by the wild/weedy plant. Therefore, the genetic makeup—and hence the appearance—of the introgressed plant is primarily that of the wild/weedy plant (see "Becoming Wild" in chapter 4). After several generations of repeated backcrosses, only select genes from the crop will be retained by wild/weedy plant populations. But the transgene may be among those retained in the populations. In fact, although only a few crop genes may remain in a population following transfer in any one growing season, the presentation of the full suite of crop genes to wild/weedy populations season after season virtually guarantees a flow of transgenic crop genes to wild/weedy populations.

Factors Influencing Gene Flow

The significance of gene flow is determined by the extent to which (i) crosses between crops and wild/weedy relatives produce fertile hybrids in which the gene(s) are expressed and (ii) the transgenes are retained in a population (Caplan and Montagu, 1990; Crawley, 1990a; Doebley, 1990; Ellstrand and Hoffman, 1990; Hoffman, 1990; Keeler and Turner, 1991; Manasse and Kareiva, 1991; National Research Council, 1989b; Regal, 1988; Wilson, 1990).

Sexual Reproduction in Flowering Plants

Sexual reproduction occurs among crops, weeds, and wild plants within the distance that pollination can occur and where the plants are sexually compatible. Because pollen transfer is mediated primarily by wind and insects, the distance within which pollination can occur is affected by wind turbulence, speed, and direction and/or the flying range of insects (Regal, 1982). Another factor is the longevity of pollen itself. Generally, pollen is viable only a short time; that is, it is capable of fertilizing eggs only within a time period of hours or days, not weeks, after it is produced.

Plants that are sexually compatible can successfully reproduce together sexually. This means that an egg in one plant can be fertilized by pollen from another and that the fertilized eggs develop into viable, fertile offspring (figure 3.1). Populations may vary in their degree of compatibility. If populations A and B show a greater degree of sexual compatibility than C and D, then pollination between A and B will more often result in production of viable seed than pollination between C and D. Some plants are self-compatible; that is, the pollen of a plant can fertilize the eggs of the same plant or other plants of the same variety. Plants may also be self-incompatible, a means of forcing outcrossing with other plants.

Sexually compatible plants are closely related and are typically classified in the same species. Organisms in the same genus but of different species may occasionally interbreed as well. In many cases, crops and their wild relatives have been given different species names even though they are sexually compatible. Examples are *Pisum sativum* (pea) and its compatible wild relative *Pisum humile*, and *Lens culinaris* (lentil) and its compatible relative *Lens orientalis* (J. Hancock, personal communication, 1992). Occasionally, plants not classified as closely related may interbreed. Examples include interbreeding among many cacti and orchids (N. Ellstrand, personal communication, 1992) and between wheat (*Triticum aestivum*) and a weed, jointed goatgrass (*Aegilops cylindrica*), in a different genus (Callihan et al., 1990; Donald and Ogg, 1991). [Some taxonomists place wheat and jointed goatgrass in the same genus, *Triticum* (Bowden, 1959; Gould and Shaw, 1968)].

Sexual reproduction in plants is initiated when pollen from one plant lands on the female part of a flower of a sexually compatible plant (figure 3.1). Pollen grains produce tubes that grow into the female part of the flower and from which the plant's eggs are fertilized. The fertilized eggs, containing genes from the pollen and the egg, grow into embryos within seeds. Where these seeds contain genetic contributions from two varieties of parental plants that differ in one or more traits, they are called hybrids.

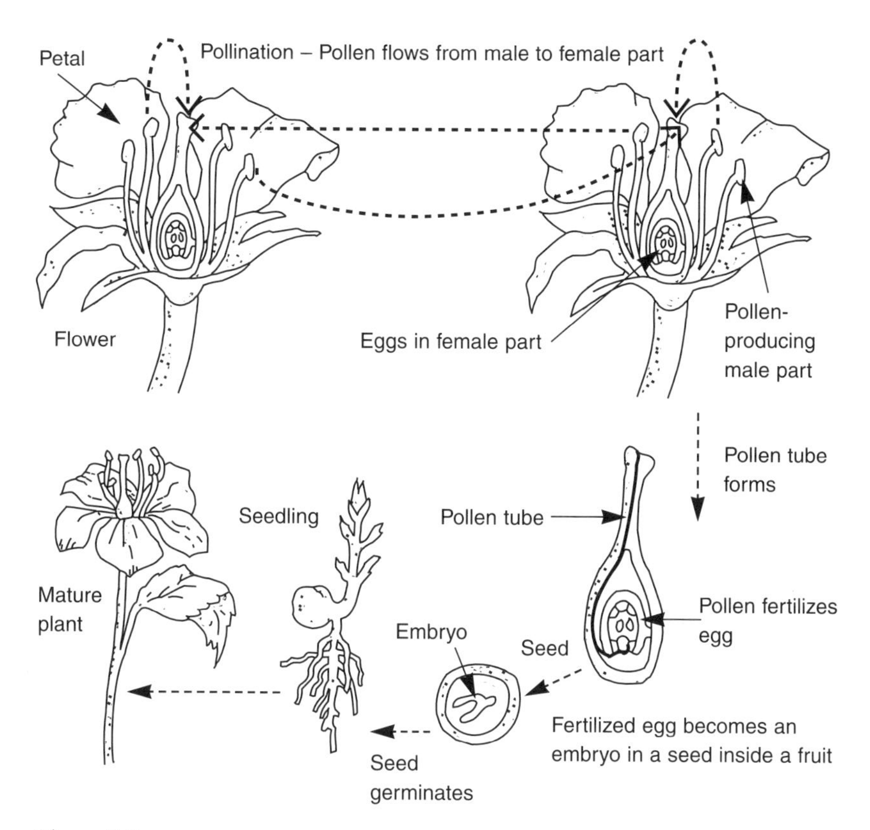

Figure 3.1
Sexual reproduction in flowering plants

Keeler and Turner (1991) describe factors that influence the likelihood of successful crosses between crops and relatives. First, crops and relatives must grow in proximity to one another; that is, within the distance that viable pollen is dispersed. In addition, the relative and the crop must be compatible and must cross and produce viable, fertile offspring in which the transgene is expressed.

Proximity of Compatible Relatives. Crops vary in the extent to which they are planted in proximity to native and introduced wild and weedy relatives (Ellstrand and Hoffman, 1990). Moreover, the amount of empirical data defining the sexual compatibility of crops and relatives varies widely from crop to crop and relative to relative.

Table 3.3
Selected crops with and without wild/weedy relatives[a] in the United States

Relatives in the United States					
Fiber crops					
flax	cotton				
Fruit/nut crops					
blueberry	cranberry	blackberry	raspberry	plum	strawberry
cherry	peach	pecan	walnut		
Grasses, cereals, and cereal-like crops					
wheat	oats	sorghum	bluegrass	rye	millet
rice	sugarcane	fescue	ryegrass	barley	amaranth
quinoa					
Legumes					
pea	alfalfa				
Oil crops					
sunflower	peanut	rapeseed (canola)			
Vegetable/herb crops					
cabbage	broccoli	radish	turnip	cauliflower	artichoke
lettuce	squash	pumpkin	cucumber	melon	spinach
asparagus	potato	sweet potato	tomato	eggplant	pepper
carrot	celery	parsley	cilantro		
Miscellaneous					
mustard	tobacco	sugarbeet			
No relatives in the United States					
corn	soybean	dry bean	citrus crops		

a. Relatives are either in the same species or genus as the related crop. Because of a lack of uniformly reliable data on the sexual compatibility of many crops and their wild/weedy relatives, no attempt was made to indicate the extent to which interbreeding between crops and relatives produces viable, fertile hybrids, if any. Compatibility will vary considerably from crop to crop and relative to relative. In some cases, such as with the potato, evidence is accumulating that interbreeding does not occur. Some crops have only one or two wild/weedy relatives; others have many.

Sources: Boyce Thompson Institute, 1987; Council for Agricultural Science and Technology, 1991; Fuller and McClintock, 1986; Keeler and Turner, 1991; Robbins et al., 1970; Simmonds, 1976; N. Ellstrand, J. Hancock, M. L. Roush, H. Wilson, personal communications, 1992.

Table 3.3 lists crops with and without wild/weedy relatives in the United States, but does not indicate sexual compatibility. Corn and soybeans are examples of crops that have no sexually compatible wild/weedy relatives in the United States. For these crops, the likelihood of gene flow into wild populations as a result of use in the United States is virtually nil.[11] However, growing transgenic corn and soybeans elsewhere in the world, particularly in areas where wild/weedy relatives are abundant, such as the centers of diversity and origin of the crop, would almost certainly lead to flow of transgenes to wild relatives.

A number of U.S. crops are grown in close proximity to sexually compatible wild relatives. In this category in the United States are plants such as carrot, sunflower, radish, and squash. In these situations, if crop pollen is released at the same time that the compatible relatives are blooming, transgenes will almost certainly enter the wild/weedy population.

Between these two extremes are crops that are grown near wild/weedy plants that are not close relatives but may have some degree of cross compatibility. This is the case, for example, with crops in the nightshade family (*Solanaceae*)—tomato, tobacco, eggplant, pepper—which grow in proximity to some wild/weedy relatives with which the crops may or may not be compatible.

In some situations, particularly for the major U.S. crops, relevant information on sexual compatibility may be available in the plant breeding literature.[12] For many other crops, however, the level of sexual compatibility is unknown. The paucity of information may be particularly acute in some parts of the world where most of the native wild plants remain uncatalogued and certainly unstudied in regard to sexual compatibility with crops. In the United States, too, a considerable amount of information is lacking on the identity, distribution, and sexual compatibility of native and introduced relatives of many crops (Boyce Thompson Institute, 1987; Wilson, 1990).

Retention of Transgenes. Transgenes most likely to be retained in a population of wild/weedy relatives are those that enhance fitness—that is, those that confer an advantage strong enough to promote the survival of the transgenic plant over other competitors in the wild/weedy population. Thus far, transgenes engineered into crops differ widely in their likely adaptive advantage. For example, genes that provide tolerance to environmental stresses—pathogens, insects, temperature extremes, and drought—are thought to be more likely to have an adaptive advantage than genes that alter nutrient composition or confer male sterility. But since wild/weedy plants exist in varied environments under a range of climatic and biological conditions, it is difficult to predict which traits will provide the competitive edge in a particular environment for any specific population.

Neutral transgenes—those that are neither beneficial nor detrimental to the plant—may also become established and spread under three other kinds of circumstances: (i) high rate of gene migration; (ii) genetic drift in small populations; and (iii) linkage on a chromosome to other genes that confer an advantage.

A neutral transgene may be retained simply by swamping, where there are repeated high levels of gene flow (Ellstrand and Marshall, 1985; R. Linder, personal communication, 1993). Genetic drift may also operate to retain a neutral gene (Ellstrand et al., 1989 and references therein). Genetic drift refers to changes in gene frequency that arise from random events. In small populations, gene frequencies fluctuate by chance: the smaller the population, the greater the magnitude of the fluctuations (Curtis, 1983). Also, a neutral transgene may "hitchhike" along with an advantageous crop gene when the two are close together (linked) on a chromosome.

Transgenes may disappear from a population if they do not confer an ecological advantage. They may also be lost through genetic drift by the same process that retains neutral genes. For example, if a population is so small that only a single individual carries a transgene and that individual fails to reproduce, then the transgene would be lost from the population. Finally, transgenes may be retained, even if they are disadvantageous, because they are recessive (Curtis, 1983), due to genetic drift, or because of a high rate of gene flow.

Potential Adverse Impacts

Transgenic wild/weedy plants present risks similar in kind to those posed by the transgenic crops themselves. They may exacerbate weed problems on a farm, create new or worse weeds that persist in nonfarm habitats, and cause nontarget, cascading, and cumulative effects. These impacts may alter habitats, community structure, and food chain composition—ultimately affecting genetic and biological diversity. Transgenic wild/weedy plants may, in fact, turn out to be more problematic than the transgenic crop because wild relatives initially are not ecologically crippled as are some crops.

Transgenic Wild/Weedy Plants May Become Weeds in Agricultural and Nonagricultural Ecosystems. Transgenes may enhance the weediness of an already existing pest or may convert an unimportant weed to an important one. For example, the transfer of a disease-resistance gene to a minor weed whose population may have been controlled by plant pathogens could be relieved of that constraint and become a major weed problem. Or the flow of herbicide-tolerance genes to weedy relatives may exacerbate pest problems. For example, herbicide-tolerance genes transferred from a transgenic sorghum to its noxious close relative, Johnson grass, could make the weed even more difficult to control than it is now. Furthermore, the farmer would lose the economic advantage of applying the herbicide to which the sorghum was made tolerant.

The botanical and agronomic literature provides precedent for these concerns. In fact, a number of troublesome weeds have resulted from gene flow. For example, wild populations of radish (*Raphanus sativus*) have become weeds in California as a result of hybridization between escaped cultivated radish (*R. sativus*) and jointed charlock (*R. raphanistrum*)—both originally introduced from Europe in the nineteenth century (Panestos and Baker, 1967). Similarly, Darmency (1994) and Till-Bottraud et al. (1992) report that hybridization between a crop and its wild relative has produced a new species or species variant that is a more troublesome weed than the wild parent. Giant green foxtail (*Setaria viridis* var. *major*) is apparently the stable result of a cross between a crop, foxtail millet or Italian millet (*S. italica*) and a relative, wild green foxtail (*S. viridis*) (Darmency, 1994; Till-Bottraud et al., 1992).

Barrett (1983) surveyed situations where gene flow from nontransgenic crops to weeds led to the evolution of weeds that mimicked crops (mimetic weeds). Examples include rice, corn, sorghum, and millet. Because the weeds are so similar to the crop plants, they thrive under regimes meant to encourage the crops and are difficult to differentiate from crops for control purposes. In many areas of the world, for example, wild rices are serious weed problems in rice fields. Several wild rices have hybridized with cultivated rices resulting in mimetic weeds difficult to distinguish visually from the intended crop, making hand or chemical removal difficult. Rice breeders in India developed purple varieties of rice to differentiate them from the wild green rice. Soon, however, gene flow from the crop to wild rice and selection pressure from weeding produced a purple wild rice variety (Barrett, 1983 and references therein).

In some areas of Central America and Mexico, hybridization between corn and teosinte,[13] a close relative common to corn fields, produces teosinte weeds that mimic the crop. For example, Wilkes (1977) describes one form of teosinte that has become a mimetic weed by hybridizing with a particular cultivar of corn in the Valley of Mexico. The teosinte has developed the red plant color, hairy leaf sheaths, and wide leaves typical of the corn cultivar. As a result, the teosinte is difficult to distinguish from corn and escapes weeding (Barrett, 1983 and references therein).

Similarly, hybridization between the crop sorghum (*Sorghum bicolor*) and Johnson grass (*S. halepense*) produces aggressive weedy types that are vegetatively vigorous but produce little seed (Baker, 1972). The weedy hybrid persists as a perennial in sorghum fields and is difficult to eradicate because it is so similar to the crop.

Shibra (*Pennisetum americanum* ssp. *stenostacgyum*), a mimetic weed of cultivated pearl millet fields in Africa, is a result of hybridization between the crop (*P. a.* ssp. *americanum*) and the wild progenitor of cultivated millet (*P. a.* ssp. *monodii*) (Brunken et al., 1977). An additional factor in the evolution of shibra is the selection pressure for mimetic weeds due to humans weeding millet fields (Brunken et al., 1977).

Finally, weedy sugar beets provide an interesting example of a troublesome weed that has developed as a result of gene flow from a weed into a crop. In some parts of Europe, hybridization between sugar beet (*Beta vulgaris* ssp. *vulgaris*) and wild beet (*B. vulgaris* ssp. unknown) has transferred the dominant bolting trait from the weed into the crop—creating a weed problem in the sugar beet fields. Unlike the crop, the crop-weed hybrids with the bolting trait flower in their first year, making them unusable for commercial production of sugar (Dale 1994; Gliddon, 1994).

Transgenes, too, may facilitate the development of mimetic weeds. For example, the flow of herbicide-tolerance genes from a crop to weeds could render the weeds resistant to the same herbicide to which the crop plants are tolerant. That herbicide then would be useless in controlling the weeds in that crop. For many farmers, it might then be difficult to develop a new weed control strategy that avoids using multiple herbicides or avoids a herbicide that also affects the crop plant as well.

Transgenic wild/weedy plants may also become weeds in nonfarm habitats. Many nontransgenic plants have a record of invasion and successful establishment in nonagricultural systems. Their establishment generally has meant that the invaded ecosystems were changed profoundly, perhaps irrevocably. Thus, one can readily speculate that wild/weedy populations with a biological advantage conferred by transgenes could theoretically thrive in new habitats—with potential repercussions throughout the invaded community. For example, salt and cold tolerance may easily expand the range of environments in which transgenic populations can thrive.

Other Aspects of Weediness of Transgenic Wild/Weedy Plants. In addition to the creation of weeds that initially establish in fields or nonfarm habitats, transgenes flowing from crops to wild/weedy populations may set in motion a number of other potential adverse effects. Cascading and cumulative effects may be expressed in a variety of ways—reduced biological diversity in local communities and centers of crop diversity, extinction of small populations, transgene flow into subsequent crops, and effects on nontarget organisms.

Cumulative and Cascading Effects. Wild plant populations containing transgenes that confer a fitness advantage may be competitively superior to other members of the community, or they may invade new habitats—initiating chain reactions that may lead ultimately to alterations in community composition, local extinctions of plant populations, and a decline in biological diversity.

Some invading plants may displace resident plant populations, perhaps entire species. As a result, populations of other kinds of organisms may also decline: insects that feed on the displaced plants, birds that eat these insects, microorganisms that colonize the roots of the displaced plant. These shifts in plant populations may be accompanied by the introduction of new types of microorganisms, plants, insects, birds, and other animals. Other plant populations may be at an advantage in the new situation and increase in size and effect. Whatever the details of the disruption, the makeup of the original community is altered.

Several examples illustrate the impacts of nontransgenic invading plants on recipient habitats. In California, the aggressive growth of annual ryegrass (*Lolium multiflorum*), planted on burned slopes to control erosion, led to a series of events that caused a significant shift in the structure of the chaparral community native to the slopes (Mooney et al., 1986 and references therein). During a wet growing season, the ryegrass grew to a greater density than that normally produced by the native vegetation. The dense ryegrass supported a fire—a fire that destroyed seeds of the native plants. Then, a second fire killed the chaparral plants before they could reproduce and replenish the seedstocks. Consequently, the community structure was dramatically altered—with a loss of many of the native plants.

Similarly, researchers have noted that eucalyptus or blue gum (*Eucalyptus globulus*), now established as forests in California, prohibits the growth of typical native understory plants—affecting the animal species that consume them and make their homes in them (N. Ellstrand, personal communication, 1992; Thompson et al., 1987).

Purple loosestrife (*Lythrum salicaria*), a weed introduced from Europe in the nineteenth century, offers an equally strong example, because it is currently replacing native aquatic vegetation in the United States and causing subsequent changes in animal food chains and plant-microbial interactions (N. Ellstrand, personal communication, 1992). An especially successful colonizer, purple loosestrife can tolerate a wide variety of environmental conditions and soils and produces a prolific number of seeds that disperse on moving water, in mud, or even on water birds, trucks, and other vehicles. Since the 1930s, purple loosestrife has proven to be a particularly aggressive invader in floodplains of the St. Lawrence River, now posing a serious threat to native plants in marshes throughout the northeast and north central regions of the United States (Thompson et al., 1987).

> Gene flow from [transgenic] cultivars to other elements of its gene pool, both domesticated and free-living, could provide a marked selective advantage to individual recipients. . . .Progeny of these introgressed plants, armed with their unique genetic advantage, could then proceed to displace native landraces and free-living populations. . . .Thus, erosion of genetic diversity in plants, including crop/weed systems, is currently a critical problem. . .which could be exacerbated by uncontrolled application of the new technology.
>
> —H. Wilson, 1990, p. 450

Salt cedar (*Tamarix* spp.), meanwhile, shows how a weed can significantly modify its recipient ecosystem (U.S. Congress OTA, 1993; Vitousek, 1986). The tree requires wet areas for early growth, but once established it can survive by drawing water from deep roots. An extremely thirsty plant, the salt cedar absorbs large amounts of water from soil and has caused significant water loss from reservoirs in the southwest United States. It has invaded natural springs and waterways in various desert regions of the United States and, as in the Canyonlands National Park in Utah, the weed has replaced native vegetation, reduced channel width, and increased flooding (U.S. Congress OTA, 1993; Vitousek, 1986).

Marram grass (*Ammophila arenaria*), or beach grass, widely used around the world to stabilize sand dunes was first introduced to the U.S. Pacific Coast in the late nineteenth century (Slobodchikoff and Doyen, 1977 and references therein). Marram grass grows vigorously and possesses vegetative adaptations that make it successful in areas of shifting sand. It forms interlacing networks of roots that form mats under the surface, stabilizing loose sand. Dunes colonized by marram grass show reduced numbers and diversity in populations of insects as well as reduced plant diversity, compared with dunes stabilized by native plants (Slobodchikoff and Doyen, 1977 and references therein).

Engineered sunflowers provide a hypothetical example of the potential impacts of transgene flow on community composition and diversity. Transgenes for disease resistance added to sunflowers in the United States, the center of diversity of that crop, theoretically could introgress into wild populations. Then, assuming that disease is a limiting factor in sunflowers, the introgressed types—because of their greater biological advantage—may spread and displace other sunflower genetic types as well as other wild species. Not only is the integrity of natural communities altered but the options are reduced for plant breeders who turn to wild populations for new traits.

Driving Small Populations to Extinction. Small populations are particularly vulnerable to gene flow from a larger population of a related plant (Ellstrand, 1992), in this case, a crop. If a rare species receives more pollen from a crop than its own kind of pollen, either outbreeding depression or genetic assimilation (discussed below) can become an important problem. Although transgenic as well as nontransgenic crops present this threat when they are planted near small interbreeding populations, some transgenic crops may present additional complications. For example, by virtue of transgenes that confer tolerance to cold, heat, drought, or salt, some transgenic crops may be cultivated in previously inhospitable areas and may therefore cross with wild relatives never before within pollination distance of the crop.

Outbreeding depression is a "fitness reduction following hybridization" (Templeton, 1986). Depending on the species involved, hybridization can drastically reduce fitness. Fitness is lowered because the genes from each parent do not "work well" together.[14] The decreased fitness can be manifested as early as reduced seed set in the interbreeding parents or as late as the production of hybrids with reduced vigor or even later as hybrid sterility (Ellstrand, 1992; Grant, 1981; Levin, 1978). In any case, the production of a substantial fraction of unfit offspring could threaten the survival of an already endangered population.

Hybridization with crops has been implicated in the extinction of at least six wild crop relatives, including relatives of hemp, corn, pepper, and sweet pea (Small, 1984). In these cases, Small (1984) suggested that gene flow from the crop overwhelmed the small wild gene pool and gave natural populations disadvantageous genes. For example, a wild species of chili pepper (*Capsicum frutescens*) is being hybridized out of existence because of frequent interbreeding with a domesticated variety (*C. chinense*) (Eshbaugh, 1976 and references therein).

Genetic assimilation refers to the opposite problem. If hybrids are fully fertile and vigorous, the substantial pollen flow from a common crop could dilute the genetic integrity of a rare natural species until the rare form was effectively assimilated into the crop species. In essence, a rare species can be hybridized out of existence. For example, the scarce California walnut

(*Juglans hindsii*) is endangered (McGranahan et al., 1988) because hybrids between it and the common Persian walnut (*J. regia*) are typically vigorous (N. Ellstrand, personal communication, 1992).

Reintroducing Transgenes into Subsequent Crops. Another potential adverse impact, likely a rare event, concerns the capacity for a transgene to be reintroduced into subsequent crops from wild/weedy populations. This could happen when a transgene flows from a crop into wild/weedy relatives one year and then in the next season transgenes from the weedy population flow back into the new crop. Some seeds of this crop would carry the transgene. The significance of this event would depend on the nature of the transgene product, whether the transgene is expressed, and the fate of the crop, that is, whether it was to be consumed by humans or other animals or used for some other purpose.

Threatening Centers of Crop Diversity. The capacity for gene flow to alter genetic variability deserves special consideration in the case of centers of crop diversity. These biological treasures are the repositories of genes that future generations of humans will need to modify crops to meet changing environmental conditions. In Canada and the United States, for example, scientists recognize centers of diversity for berries (blueberry, cranberry, raspberry, gooseberry), sunflower, Jerusalem artichoke, pecan, black walnut, and muscadine grape (Juma, 1989; Rural Advancement Foundation International, 1992). Most other centers of diversity are in the developing world.

Despite their critical importance, the centers of diversity are losing genetic variability at an alarming rate. These losses are attributed to outright destruction, as the burgeoning human population claims more land for agriculture and habitation, and to the replacement of numerous and diverse landraces with a small number of newer varieties (Fowler and Mooney, 1990; Rural Advancement Foundation International, 1992). Abandoned landraces soon become extinct, as no one plants them or saves their seeds.

Where centers of diversity are known, countries may be able to take steps to ensure that transgenic crops are not planted within pollination distance. By contrast, centers of diversity that have not yet been identified cannot be protected. As scientists study wild crop relatives around the world, they may well identify heretofore unknown centers of crop diversity. For example, research has only recently identified the United States as a probable center of diversity for squash, in addition to the previously known centers in Central and South America (Decker, 1988; Decker-Walters, 1990; Smith et al., 1992).

Nonsexual Flow of Transgenes

Theoretically, gene flow could occur not only through the sexual mechanism discussed above but also through nonsexual mechanisms that transfer genes from one adult plant to another (Heinemann, 1991; Lewin, 1982). Such a mechanism is often called "horizontal" flow because genes move not vertically from parent to offspring but directly between adult individuals—moved perhaps by viruses or other biological vectors. Although the phenomenon is common among bacteria, horizontal gene flow from plant to plant has never been observed. Many scientists believe that horizontal transfer plays a decidedly minor role, if any, in the evolution of plants and animals, while others argue for a significant role (Regal, 1986 and references therein).

Although plant-to-plant horizontal transfer has not been documented (Hauptli et al., 1985), there are confirmed cases of transfer from bacteria to plants, viruses to animals, and bacteria to bacteria. Generally the transfer is thought to be mediated by pathogenic bacteria and viruses (Hauptli et al., 1985; Heinemann, 1991). The pathogenic bacterium *Agrobacterium tumefaciens*, for example, which can transfer genetic material from bacteria into plant cells, is widely used by genetic engineers to move genes into plants. Engineers attach foreign genes to part of the *A. tumefaciens* genome and rely on the bacteria containing the transgenes to move the foreign genes to plant cells along with a small part of the bacterial genome. The foreign genes are then incorporated in the plant cell genome. A number of animal viruses are able to transfer genes among the genomes of their animal hosts as well. Molnar and Ingram (1991), for example, suggest that viruses may be responsible for the transfer of genes between two kinds of sea urchins.

The possibility of horizontal gene transfer from plants to microorganisms remains controversial. A suggestion that an enzyme in bacteria originated from plants (Carlson and Chelm, 1986) was later refuted by Shatters and Kahn (1989) who argued that the existence of the enzyme in bacteria could be attributed to mechanisms other than horizontal gene flow. Few other claims of plant to bacteria transfer have ever been made.

Although currently no evidence exists that genes move horizontally from plants to any other organism, the phenomenon is so recently described that it is too early to conclude that transfers from plants never occur. The possibility, although remote, is important because it provides an additional avenue of gene movement. New discoveries in this area could change the assessment and significance of risk.

Summary

The flow of transgenes from crops into wild/weedy plants poses diverse ecological risks. Transgenes that confer significant biological advantages may transform wild/weedy plants into new or worse weeds that establish in farmers' fields or nearby wild habitats. Cascading repercussions of this initial transformation may ultimately mean changes in the makeup of plant communities, perhaps reduced diversity in local ecosystems, and threats to centers of diversity.

In addition, transgenes may be more problematic in wild/weedy plants than in the transgenic crop, especially where the crop has been ecologically debilitated by conventional breeding. Finally, nonsexual or horizontal gene flow mechanisms theoretically may move transgenes from one adult plant to another. However, there is no evidence that these transfers occur.

Risks to Plants from Transgenic Virus-Resistant Crops

This section considers other potential environmental impacts of transgenic crops—risks posed by certain virus-resistant crops.[15] They do not fall neatly into the gene flow categories discussed above because the foreign gene comes from an infectious agent.

Many agricultural biotechnology labs are working to engineer plants to resist the damaging effects of plant viruses. To accomplish this, engineers splice viral genes into a crop plant's genome (Beachy, 1990; Beachy et al., 1990 and references therein; Matthews, 1991 and references therein; Powell-Abel et al., 1986; Tien, 1990). The resulting transgenic plant is resistant to infection by the virus from which the viral gene was taken.

Some scientists have raised the possibilities that widespread use of transgenic virus-resistant plants in agriculture may present risks: producing new strains of viruses, allowing a virus to infect a new host, or exacerbating existing viral diseases (de Zoeten, 1991; Palukaitis, 1991; Tepfer, 1993; Tolin, 1991). These risks, if they exist, will be presented to wild/weedy plants as well as crops. Like crops, noncrop plants are susceptible to viral diseases and therefore may be affected by any new viral strains, extended host ranges, and exacerbated diseases that may arise from transgenic crops. In addition, viral transgenes may move via pollen, introgress into populations of sexually compatible wild/weedy populations, and confer viral protection.

Scientists have attempted to engineer plants for resistance to infection by incorporating genes for viral products into the plant genome. The strategy relies on the largely unexplained capacity of the viral genes or gene products to confer a protective effect once they are synthesized by the plant. Thus far, scientists have used primarily coat protein genes to produce resistance. Less frequently, they have used small pieces of ribonucleic acid (RNA) called satellite RNA and genes for other viral components.

Less than a decade ago, scientists found that the presence of a coat protein gene[16] in transgenic plant cells conferred protection against virus infection by some unknown mechanism (Beachy, 1990; Beachy et al., 1990 and references therein; Powell-Abel et al., 1986). Researchers later confirmed that transgenic coat protein-containing plants are resistant to the virus from which the coat protein was derived as well as to closely related strains of that virus (Beachy, 1990). The phenomenon, called "coat protein-mediated protection," resembles but is different from another phenomenon—cross protection[17]—known for many years to occur naturally in plants. The mechanism by which coat protein genes confer protection against virus infection remains unknown.

Some background is important in understanding the possible risks associated with the expression of coat protein genes in transgenic plants. Field-grown crops are often simultaneously infected by several viruses—the particular ones vary according to crop, availability of insect vectors,[18] and environmental conditions. Thus, transgenic crops may contain several viral populations different from the one that is the source of the coat protein gene. In some cases, the coat protein of one virus may completely or partly enclose the genome of another virus—a rare phenomenon termed transcapsidation.[19] The coat protein can determine which insect transmits a virus (Pirone, 1991; Rochow, 1977).

In addition to coat protein, other virus-associated molecules can be used to confer resistance on a plant. Scientists have successfully conferred protection by engineering plants to express so-called satellite RNAs (satRNAs) (Harrison et al., 1987; Tepfer, 1993 and references therein; Tien and Wu, 1991 and references therein). For example, Harrison et al. (1987) found that cucumber mosaic virus (CMV) replication[20] was significantly decreased and symptoms mostly suppressed in satRNA-transgenic plants compared to nontransgenic plants. McGarvey et al. (1990) reported that satRNA-transgenic tomatoes developed resistance to some CMV strains. Like coat protein-mediated protection, the mechanism underlying this phenomenon is not known.

SatRNA refers to small pieces of RNA that are occasionally associated with some strains of certain plant viruses. For example, isolates of CMV and tobacco ringspot virus may have associated satRNA. The virus with which the satRNA is associated is called the "helper" virus. Specific satRNAs are associated with specific helper viruses. SatRNA and its helper virus may interact in a number of ways. The satRNA, a separate piece of RNA replicated on its own, depends on the helper virus for successfully making copies of itself. In turn, satRNA disrupts, to some extent, the normal replication of the helper virus (Matthews, 1991). The satRNA, with no genetic code for its own coat protein, is enclosed in the coat protein of the helper virus, alone or along with, the helper viral RNA (Collmer and Howell, 1992 and references therein). SatRNA can cause disease only in association with the helper. By contrast, the helper virus can cause disease alone.

In some hosts the satRNA along with the helper virus may cause a new disease (Collmer and Howell, 1992 and references therein). A satRNA can also act as a disease regulator in plants infected with a helper virus, reducing or less commonly, exacerbating disease symptoms (Collmer and Howell, 1992 and references therein; Waterworth et al., 1979).

In transgenic plants where satRNA is incorporated into the plant genome, the satRNA is replicated and available for encapsidation[21] in helper coat protein if the cell is infected by the helper virus. The encapsidated satRNA can then be transmitted by vectors of the helper virus (Boyce Thompson Institute, 1987; Harrison et al., 1987; Palukaitis, 1991).

Other factors that are relevant to the later discussion of risks are: (i) satRNA may be benign in the transgenic plant and virulent in a different plant species; (ii) differences between benign and harmful satRNAs may involve a change in only one or a few nucleotides (Gordon and Symons, 1983; Sleat and Palukaitis, 1990, 1992); and iii) satRNA has a high mutation rate (Boyce Thompson Institute, 1987; Kurath and Palukaitis, 1990).

Work is also under way in a number of laboratories to confer resistance by using other components of the viral genome, such as genes for so-called nonstructural proteins[22] (for example, Golemboski et al., 1990; Matthews, 1991 and references therein).

New Viral Strains May Arise

De Zoeten (1991), Tepfer (1993), and Tolin (1991) suggest that new viral strains could arise from recombination, the exchange of pieces of nucleic acid between two similar nucleic acids resulting in new arrangements of genetic material. In this case, it is proposed that exchanges may occur, albeit rarely, in a transgenic plant between the nucleic acid of the coat protein gene and the nucleic acid of a related virus. After the exchanges, the recombinant virus has a different genetic makeup and in some cases altered biological characteristics. For recombination to occur, viral nucleic acid and the viral transgene must be in the same place at the same time and be similar enough to allow exchanges to occur. Although researchers have only just begun addressing the ecological implications of recombination, some scientists speculate that recombination may play a role in the evolution and survival of plant viruses or, under certain conditions, produce a new viral strain with an altered host range (Allison et al., 1990; Bujarski and Kaesberg, 1986).

Recombination has been observed in transgenic plants. Gal et al. (1992) showed that recombination occurs between cauliflower mosaic virus (CaMV) and a CaMV gene located on a plant chromosome of transgenic turnip. Similarly, Lommel and Xiong (1991) found recombination between red clover necrotic mosaic virus (RCNMV) and an RCNMV gene incorporated into a plant chromosome.

Another report shows that recombination can occur in a transgenic plant and new forms of a virus can arise as a result. Schoelz and Wintermantel (1993) report that cauliflower mosaic virus (CaMV) recombines with CaMV transgenes on a plant chromosome and thereby obtains sequences from the transgenic plant that alter the host range of CaMV and change the symptoms produced. In this case, the recombinant virus can systemically infect a wider host range than the CaMV that originally infected the transgenic plant. Greene and Allison (1994) also showed that recombination between a cowpea chlorotic mottle virus (CCMV) mutant strain and a CCMV transgene in a host restored the ability of the virus to cause systemic infection. Before recombination, the mutant strain was not able to systemically infect plants.

Viral Host Ranges May Be Broadened

Unlike recombination, which might create new viral strains, transcapsidation, the encapsulation of viral core with coat protein from a different virus, is typically a short-lived phenomenon. After the transcapsidated virus reaches its new host, it will replicate itself and its own coat protein once again. For viruses that are transmitted by insects and seeds, transcapsidation then raises the possibility that the range of hosts that viruses can infect may be altered, albeit transiently. The following series of events explains how an insect-transmitted virus transcapsidated in an engineered virus-resistant plant might infect a new host. Figure 3.2 illustrates transcapsidation in an engineered plant.

- A transgenic crop has been engineered to produce coat protein of virus X to protect the crop against infection by virus X. Virus X is an insect-transmitted virus.
- Virus Y infects the transgenic crop.
- Coat protein of virus X encapsidates the genome of virus Y.
- The insect that normally transmits virus X can now transmit virus Y (a Y genome enclosed fully or in part in coat protein X) to hosts to which virus Y may not have been previously transmitted but in which it can cause disease.
- Virus Y has a new host range.

The likelihood that all these events would occur in such a way to produce a virus with an expanded host range is not known and will vary

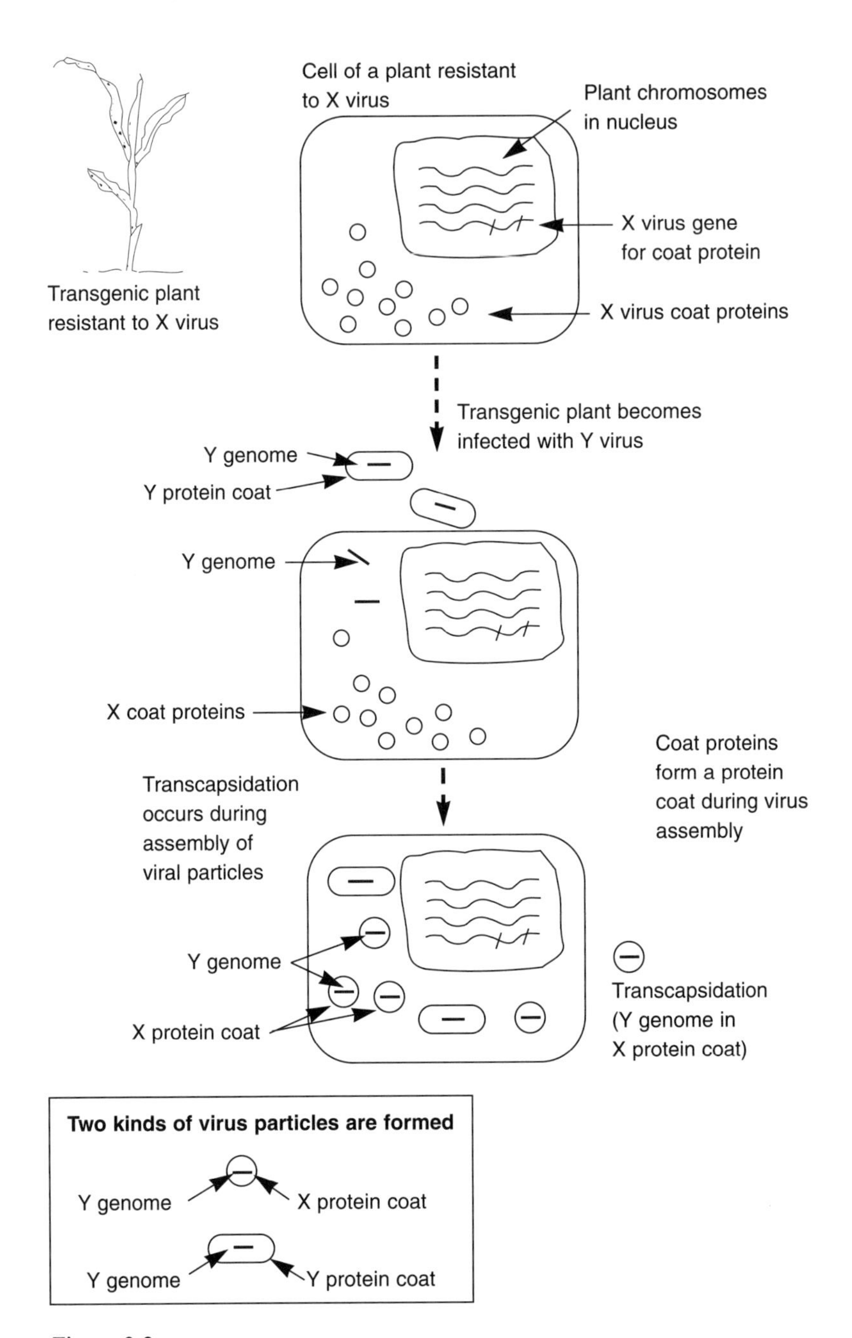

Figure 3.2
Transcapsidation of a virus in a transgenic virus-resistant plant

considerably from virus to virus and crop to crop. In fact, there are viruses for which this series of events will not occur.[23]

Recent research, however, demonstrates that the proposed chain of events could occur in certain situations. Chen and Francki (1990) reported that the coat protein of cucumber mosaic virus encapsidated the tobacco mosaic virus (TMV) genome in their experimental system. Furthermore, they showed that insects transmitted the hybrid particle to a plant that TMV subsequently infected (TMV normally is not insect transmitted).

Farinelli et al. (1992) showed that transcapsidation occurs in transgenic plants infected with a viral strain. Plants transgenic for coat protein of one strain of potato virus Y (PVY) were inoculated with a different PVY strain. Subsequently, the infected transgenic plant produced viral particles of two types—one with coat protein of the infecting strain and one with coat protein of the transgene.

Lecoq et al. (1993) have gone one step further. They have shown that transcapsidation in plants transgenic for coat protein can alter the vector transmissibility of an infecting virus. They engineered plants to contain the coat protein gene of an insect-transmissible strain of plum pox virus (PPV). Then they infected those plants with an insect-nontransmissible strain of zucchini yellow mosaic virus (ZYMV). A defective coat protein was responsible for the nontransmissibility of the ZYMV strain. Subsequently, the researchers found viral particles composed of ZYMV genome inside both PPV and ZYMV coat protein. Unlike the original ZYMV strain, the resulting ZYMV was insect transmissible.

As mentioned above, an expanded host range due to transcapsidation, if it occurs at all, is generally expected to be a "one-time" phenomenon. Once virus Y is transmitted to a new host that its natural vector would not feed on, it will replicate in the plant only if it is compatible with the plant cell's molecular machinery. If this happens, virus Y will be encapsidated in its own natural coat protein since it does not possess the genes to make the coat protein (X) with which it came to the plant. This virus then will be transmitted by its natural vector to its natural host, not the new host to which it was taken with the X coat protein. Thus, the transcapsidation produces a temporary extension of the host range.

Even the transient extension, however, may have economic implications if the new host is severely damaged by the virus that heretofore was not delivered to the crop. In addition, the vector may carry the transcapsidated virus outside the field to other plants that may for the first time support infection by virus Y. A rare event may become important if it occurs many times in a growing season amid large acreages of transgenic crops and potential new hosts of transcapsidated viruses.

Also, there is at least one scenario under which the phenomenon would not be temporary. If a number of different agricultural crops were transgenic for the same coat protein X and if virus Y were transmitted to those plants, then arriving virus Y would continue to be provided with coat protein X and could subsequently be transmitted to additional plants (P. Palukaitis, personal communication, 1992).

For viruses that are transmitted through seed,[24] Palukaitis (1991) raises the question of whether transgenic plants expressing coat protein show any increase in the level of seed transmission of transcapsidated viruses. In other words, does transcapsidation make it possible for some viruses, which naturally are not seed transmitted, to be seed transmitted? Or does transcapsidation enhance the seed transmissibility of viruses that are already seed transmitted?

For viruses that are already seed transmitted, enhanced seed transmissibility means that more plants of the next generation in a field will be infected with the virus,[25] thereby increasing the reservoir of viruses that can be then carried to other plants in the field. For viruses not previously transmitted by seed, the new trait of seed transmissibility opens up a new source of virus for subsequent infection in the field. This phenomenon would be a transient one unless the transcapsidated virus is moved to another transgenic plant where it can obtain coat protein.

Palukaitis (1991 and references therein) offers the following example to illustrate the potential problem. CMV is seed-transmissible to varying degrees in a number of plant types. That is, in some crops under certain conditions, CMV may be transmitted at a fairly high rate (for example, 50 percent) through seed. In other situations, the rate may be less than 10 percent. The question is whether CMV coat protein that encapsidates the other viruses infecting the transgenic plant will alter the seed transmissibility of the other viruses (which may not be seed transmissible) and thus increase rates of infection in crops where the viruses are normally low-level problems or extend the pathogens to new types of crops.

The potential for transcapsidation in transgenic plants to alter seed transmission can be evaluated by comparing the levels of infected seed from transgenic plants inoculated with a virus that could be transcapsidated with seed from nontransgenic plants similarly inoculated.

Existing Viral Diseases May Be More Severe

Another concern is that interactions between the transgene products and other viruses or viral products may enhance disease development, seed transmission, or other disease-related phenomena. For example, large acreages of plants engineered to express satRNA mean greater amounts of satRNA produced in plant cells that are available for replication and encapsidation along with any number of helper virus strains naturally infecting the crop. Where satRNAs and helper viruses interact to cause a new or exacerbated disease, the increased amount of satRNA available as a result of transgenic crops may increase the incidence of the diseases resulting from the interaction. In addition, frequently mutating satRNAs may change from benign to pathogenic in the transgenic source plant (Harrison et al., 1987).

Another possibility is suggested by Palukaitis (1991): transgenic viral products may interact with other viruses to cause more severe diseases. Known as "viral synergism," this phenomenon occurs where viruses interact to cause a more severe disease than the additive effects of the two viruses simultaneously infecting the plant[26] (Matthews, 1981).

Since the role of coat protein and satellites in synergism is unknown, it is not clear what effect coat protein and satRNA transgenic plants would have, if any, on synergistic interactions with other viruses naturally infecting the plants in the field (Palukaitis, 1991). If coat protein, for example, were involved in synergism with another virus, plants transgenic for coat protein would suffer a more severe disease when infected by that virus than would nontransgenic plants. Whether viral synergism is likely to occur in transgenic plants can be evaluated by inoculating coat protein-transgenic plants with viruses that normally infect the crop and comparing the results with nontransgenic crops inoculated with the same viruses (G. de Zoeten, personal communication, 1992).

Finally, work in other laboratories raises the possibility that interactions in infected transgenic, virus-resistant plants may temporarily convert a mild or defective[27] virus into a more severe strain. For example, Osbourn et al. (1990) showed that two defective viral strains could cause systemic infections in a plant transgenic for the coat protein of the virus. This group worked with two mutants of TMV which were defective in their ability to spread throughout an infected individual tobacco plant, whereas they could cause localized damage. However, these mutants, when grown in transgenic tobacco plants that produced TMV coat protein, were no longer defective in spreading ability. In other words, the mutants could now spread through the plant causing systemic infection.

Schoelz et al. (1991) showed a similar conversion of a defective strain of CaMV into a systemically infecting one when it was placed in a plant transgenic for the piece of virus missing from the defective strain. A particular viral gene, numbered VI, allows CaMV to systemically infect some hosts. The scientists engineered plants to contain the VI gene. Then they infected the transgenic plants with a CaMV strain defective in the ability to systemically infect a host. Their results showed that the defective strain was converted to a competent one in the presence of the transgene VI product. This phenomenon would be transient because it could occur only in transgenic plants with the complementary viral product. Once the virus infected a nontransgenic plant, it would revert to its defective state because it does not code for the full set of gene products.

Summary

This section has discussed risks—new viral strains, altered host range and more severe diseases—that may arise from widespread use of some engineered virus-resistant crops. Some scientists (for example, Asgrow Seed Company, 1992; Falk and Breuning, 1994) contend that these risks can be ignored because natural coinfection of plants by multiple viruses can have similar effects, presumably at similar rates. This claim can and should be investigated experimentally—-by comparing the rates and impacts of recombination and transcapsidation in engineered and nonengineered crops. If indeed the risks of transgenics are shown to be comparable to natural coinfection, transgenic virus-resistant crops would not require additional assessment or control.

Unknown Risks

The discussion above provides a comprehensive and inclusive catalog of the potential adverse effects of broad-scale use of transgenic plants as they are currently perceived. But in a world all too often confronted with scenarios like loss of the ozone layer and global warming—both unforeseen effects of earlier technologies—the question of unknown risks must be raised.

The prediction of consequences of a technology in advance of its implementation is inherently difficult. The difficulty is enhanced where the technology, as here, encompasses broad and varied activities—involving many crops, from sunflower to corn, that are transformed with hundreds of genes with the potential to spread into millions of plant populations in different environments. Under these circumstances, it is important to acknowledge that the catalog of risks—even the most complete available—may have missed something.

Some risks may be missed simply because the understanding of physiology, genetics, and evolution, among other disciplines, is limited. What such risks might be are, by definition, hard to imagine. Indeed, the set of concerns with virus-resistant crops could hardly have been anticipated five to ten years ago, since the protection of plants with viral genes was largely an empirical finding—and the mechanisms are still not understood. This reinforces the conclusion that risk assessment of transgenic plants is still in its infancy as a science.

There are sufficient uncertainties in our understanding of mechanisms of evolution, for example, that unexpected effects of new genes in a gene pool cannot be ruled out. As an illustration, evolution may have relied on a very small proportion of available DNA changed in time only slowly through mechanisms like gene duplication. Direct transfusions of new functional DNA into plants may indeed be utterly new from an evolutionary standpoint. Or perhaps large evolutionary steps, like speciation, may depend on mobile elements that engineers may be moving, activating, or suppressing without knowing their importance.

Such speculations cannot be incorporated into crop-by-crop assessments. But they serve as reminders that risk assessments are always limited by the questions that one can think to ask; that the understanding of genetic control of development is at a rudimentary stage; and that genetic engineering is far from the precise applied discipline that the use of the term engineering implies.

Summary

This chapter has discussed the two broad categories of ecological risks presented by commercial uses of transgenic crops: risks from the transgenic crops themselves and from wild/weedy plants that may be the recipients of genes flowing from the transgenic crop.

Transgenic plants may become weeds with consequent diverse effects in agricultural and nonfarm ecosystems. In some cases, the establishment of new weeds could set off a series of events that may ultimately accumulate as impacts on structures of plant communities and food chains or as new problems in farm management.

Transgenic virus-resistant plants may broaden the host range of some viruses or allow the production of new virus strains. The likelihood of these effects ranges from zero to significant depending on the situation. But generally, we believe that virus-related risks are likely to be of less concern than those associated with weediness.

Unfortunately, only two aspects of the risk of transgenic crops are currently amenable to practical assessment: the capacity for transgenic plants to invade ecosystems and the likelihood that genes may flow from transgenic crops to wild/weedy relatives. The next chapter develops a scheme for obtaining empirical data to assess these two aspects of ecological risk.

4

Two Risk Scenarios: An Experimental Assessment

This chapter offers a new framework for experimentally assessing two major environmental risks identified and discussed in chapter 3: the weediness potential of transgenic plants and the implications of transgene flow to relatives of crops.

Assessing all the aspects of weediness outlined in chapter 3 is impossible at current levels of scientific understanding. To develop a practical risk assessment of weediness potential, this chapter focuses on an initial step in becoming a weed, the capacity to invade ecosystems. Where this chapter uses the terms "weeds" or "weediness," it refers primarily to invasiveness.

Introduction

Risk Assessment

In the United States, regulatory agencies have varying degrees of responsibility and authority to restrict actions that threaten human health and the environment. To protect against possible harm, agencies must be able to distinguish risky activities from safe ones. In the case of engineered crops, regulatory decisionmakers will rely on risk assessments to gauge the likelihood that a particular transgenic crop will cause harm to public or environmental health. Using such assessments, agencies attempt to predict the behavior of the novel plants in ecological and agricultural contexts before they are released.

A risk assessment about a new engineered crop contains several interrelated judgments and determinations. These include: (i) the severity of potential adverse impacts; (ii) the potential exposures of humans or other elements of the environment to the crop; (iii) the degree of uncertainty inherent in the analysis because of gaps in information; and (iv) the capacity to mitigate the risk, that is, to control the crop once it is released.[1] Such an analysis is used to arrive at an overall likelihood that a crop will cause harm. Generally, an assessment is based on existing information as well as on new data that must be generated.

The fundamental question addressed in this chapter is whether the presence of a transgene alters recipient plants in ways that make them a new or worse weed compared with nontransgenics. One could approach this question in a number of ways. The strategy offered here is a compromise between two extremes that differ markedly in their dependence on empirical data.

At one end of the spectrum, a risk assessment might rely on abstract arguments based on what people think a particular transgene might do in a particular crop or in a particular wild relative. This kind of assessment, then, would rely primarily on existing knowledge about a particular crop and a transgene. Predictions about weediness of a plant, for example, might be based on the presence or absence of traits often associated with weediness (Keeler, 1989; Perrins et al., 1992). At the other end of the spectrum, an assessment would rest on an evaluation of observed behavior in the field. To be completely reliable, such an assessment could require extensive experimentation examining a full range of traits in many environments over a period of many years.

The approach offered here is a compromise between these two. It uses existing information, but also relies on some relatively simple experiments to predict risk. This proposal requires that all transgenic crops be subject to experimental analysis, at least until a body of data is collected on the behavior of transgenic crops in the environment. But it does differentiate among crops in setting experimental requirements. Based on existing information, crops that appear to present a higher risk of weediness are subject to more experimentation than lower-risk ones.

This approach involves a three-tiered analysis to evaluate both crop weediness and gene flow. The tiers are designed to identify nonrisky plants early in the analysis and to require extensive field testing only for plants that appear to pose substantial risks. In the case of both crop weediness and gene flow, the first tier assessment is based, for the most part, on existing data. The second and third tiers, which are essentially the same for crops and wild/weedy plants, require that experimental data be generated on the plants themselves.

By necessity, the proposed scheme is described in general terms because of the important differences among parent crops, transgenes, and growing environments. Moreover, the general analysis that makes up the three tiers should not be seen as rigid and unchanging. Rather, at some points, for some crops, special considerations, as illustrated below, may be brought to bear on the assessment.

The approach offered here is intended to stimulate discussion on ways to assess weediness and gene flow. (See also Linder and Schmitt, 1994.) As experience is gained and research is conducted, this assessment strategy will be modified, and others will be developed to reflect the new knowledge.

Population Replacement Experiments to Predict Risk

The proposal outlined in this book relies on empirical data from relatively simple population replacement experiments to indicate the potential for transgenic plants to become weeds. Population replacement refers to the capacity over time of a population of plants to produce viable, fertile offspring. A plant population can be replaced from seeds that germinate the next year[2] as well as viable seeds that remain in the seed bank.[3]

Population replacement is a measure of the increase or decline of a genetic type over generations—that is, whether a particular genetic type will persist over time. This measure encompasses two pieces of information: the rate at which a population replaces itself and the persistence of its seed bank.

Data from population replacement experiments are used to compare the "ecological performance" of transgenic versus nontransgenic plants. For example, if the experiments show that the population of engineered plants declines and its seed bank persists less well relative to the nontransformed counterpart, then it is highly unlikely that the transgenic plant will be able to cause greater adverse impacts than the nontransgenic plant. Generally, an organism that is maintaining itself less well in the environment than a counterpart is would not be expected to pose a greater risk than that counterpart over time.

Experiments measuring population replacement are admittedly an imperfect predictor of the risk posed by plants, however. They are limited because empirical data can be collected for only a few generations in a few kinds of environments. Nevertheless, assessments based on actual performance are superior to those based only on abstract considerations of characteristics thought to contribute to risk. In this book, the term "ecological performance" will be used to describe the overall results of population replacement experiments for a particular crop.

The proposed assessment plan employs tiered decisionmaking schemes involving population replacement experiments to evaluate both the transgenic crop and the transgenic wild/weedy plants created through gene flow. The population replacement experiments fall in the second tier—between an analysis (tier 1) relying on existing data and a more extensive experimental assessment (tier 3) of the few crops that reach this stage.

Assessing the Potential for Transgenic Crops to Become Weeds

This section proposes a scheme under which all transgenic plants would be evaluated for their potential to become weeds. Under a tiered analytical strategy, the first decision point would rely on existing information to place each transgenic crop into one of two groups that differ in risk or the potential for weediness. In tier 2, both groups are then subject to field experiments that evaluate ecological performance. The experiments differ in extent—corresponding to the level of risk. The risk assessment ends at the second tier for transgenic crops performing no better than their nonengineered counterpart, as they are not likely to become worse weeds than the nontransgenic form.

Transgenic crops showing enhanced performance proceed to the third tier of testing, if the company chooses to continue the risk assessment. The third tier requires experimental evaluation of many traits associated with weediness potential. Figure 4.1 summarizes this general scheme. It should be noted that special considerations not included in the generalized assessment may enter the analysis of some crops.

In the United States, regulatory agencies may eventually conduct tier 1 analyses for every crop grown in this country. Then companies would

Tier 1: Weediness of the parent crop
Is the parent crop weedy or does it have close weedy relatives in North America?

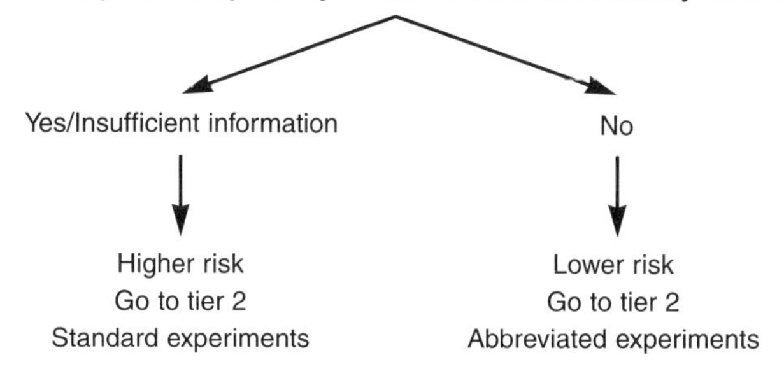

Tier 2: Ecological performance of the transgenic crop
Does the transgenic crop outperform the nontransgenic crop?

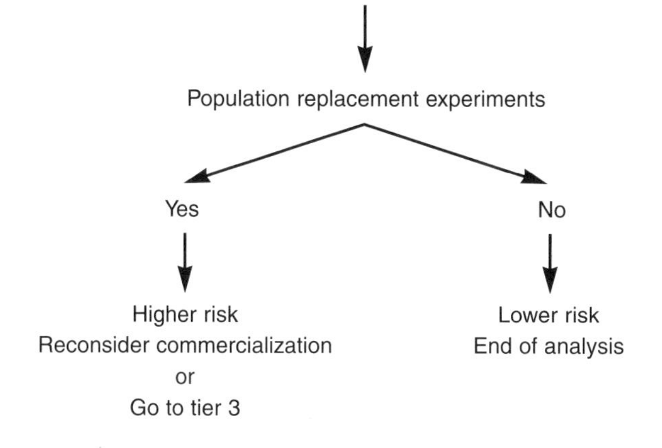

Tier 3: Weediness of the transgenic crop
Is weediness increased in the transgenic crop?

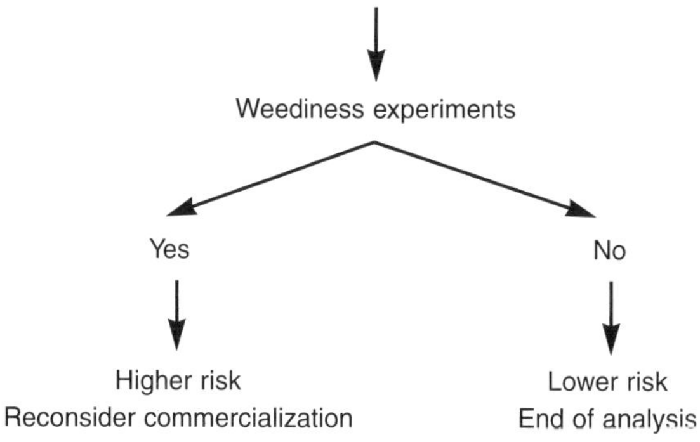

Figure 4.1
Experimental assessment of the potential for transgenic crops to become weeds

know in advance the population replacement experiments expected of their transgenic crop. Once this crop-by-crop analysis is completed, companies can eliminate the tier 1 analysis.

This scheme, requiring that all transgenic crops undergo some form of population replacement experiments, is an admittedly conservative one—based on the significant amount of uncertainty and the paucity of data associated with the environmental behavior of genetically engineered crops. Once there are ecological data on transgenic crops grown in a wide variety of environments over a period of several years, it may be possible to exempt some transgenic crops from the replacement experiments.

Tier 1. Weediness of the Parent Crop: Is the Parent Crop Weedy or Does It Have Close Weedy Relatives in North America?

The tier 1 analysis determines whether the parent crop is a weed or has close weedy relatives by searching the literature and consulting experts for information on the field behavior of the plant. If the search reveals that the crop or a near relative is a known weed, the parent would be subject to a standard set of experimental tests. If the crop is not known to be a weed or closely related to a weed, it would nevertheless be subject to field tests, although of an abbreviated nature.

The evaluation proceeds under the assumption that crops on the lower end of the spectrum of weediness potential are sufficiently unlikely to be converted to weeds by the addition of transgenes that they can be subject to simplified population replacement experiments. In other words, some crops are so genetically debilitated and dependent on humans for survival and reproduction that the addition of new traits—even advantageous ones—would not be expected to convert the crop to weeds.

Requiring these crops to be subject even to an abbreviated set of experiments is a cautious approach that provides an opportunity—one that is not unduly burdensome—to evaluate at least a small amount of data relevant to the prediction. Eventually, as data are accumulated, it may be possible to eliminate the population replacement experiments for some crops.

Using existing information, the best assessment of weediness potential of the parent crop is to ask whether the parent is a weed anywhere in a particular region. The answer reveals whether the parent possesses traits that can confer weediness under specific or broad environmental conditions. For the assessment of crops to be commercialized in the United States, North America is proposed as the region for which the question should be asked. Because U.S. ecosystems clearly cross the borders with Canada and Mexico—thereby providing opportunities for U.S. crops to be weeds or have weedy relatives in the two neighboring countries—examining parent weediness in these three countries gives a margin of safety in the risk assessment.

In some instances, weediness of a crop in the state of Hawaii or other U.S. islands may warrant special consideration. A crop that is weedy solely in Hawaii or other islands, and nowhere on the North American continent, may be placed in a lower-risk group provided the seeds of the crop will never be grown (even for test purposes or seed increases) on the islands. This consideration is justified on the grounds that the island ecosystems are sufficiently different from mainland ones that weediness solely on the islands is not a good predictor of weediness on the mainland.

At one extreme of the spectrum of parent weediness in the United States are Johnson grass and sunflower, crop plants that are well known as weeds. At the other end is soybean, a highly domesticated crop that in the United States has not become a weed despite decades of cultivation. In between, with varying degrees of weediness potential, are most of the other crops cultivated in the United States.

If the parent crop is known to be a weed, the transgenic crop is placed in the higher-risk category and subjected to the standard set of population replacement experiments in tier 2. If, on the other hand, the crop is not known to be a weed, the analysis proceeds to consider another factor that may give a clue to weediness potential, that is, whether any races of the crop or close relatives are weeds. As noted in chapter 3, having close relatives as weeds may indicate a potential for weediness in the crop. In this report, a close relative refers to a plant that is classified in the same taxonomic genus as the crop.[4]

A transgenic crop whose parent is neither weedy nor possesses close weedy relatives is placed in the lower-risk category and is subject to an abbreviated set of experiments in tier 2. If, on the other hand, a nonweedy parent crop has close weedy relatives, the transgenic crop is placed in the higher-risk category. Insufficient information to assess the weediness status of the parent crop or its relatives places that transgenic crop in the higher-risk category.

In some cases, even though a crop and weedy relatives share the same genus designation, special consideration may be given to crops that are so genetically debilitated that this criterion is not relevant. If such a case can be well substantiated, that crop can proceed to the abbreviated set of tier 2 experiments. Corn, for example, has relatives—teosinte—in Mexico that are sometimes weeds in corn fields. This finding would place corn in the higher-risk group for tier 2 testing. However, because of its well-known inability to survive without human intervention—a result of hundreds of years of breeding—this crop could be placed in the lower-risk group for tier 2 testing.

Finally, crops that do not have a long history of cultivation in a country or region should be placed in the higher-risk grouping. This conclusion is based on the assumption that the weediness status of a recently introduced parent crop may not have been determined since it will not have had sufficient opportunities to escape and become established as a weed in the new environment if it has that potential.

Sources of Information for Tier 1 Analysis. There is no central repository of information on the weediness or potential weediness of various crops and their relatives. Instead, a variety of sources may provide the information.[5] These sources, including floristic surveys, weed lists, herbaria, botanical and agronomic literature, and botanists and crop specialists, will vary in both quality and quantity from crop to crop. In some cases, new information may have to be developed.

Floristic surveys list the flora, or species of plant life (primarily flowering plants), of a given area.[6] The information, generally compiled and published by academic taxonomists, may be found in university libraries. The surveys may contain information on whether a specific plant is a weed; in particular, they may list so-called ecological weeds, that is, weeds that persist in disturbed areas, as opposed to feral weeds, plants that were once domesticated and then became wild. The information may give clues to the aggressiveness of crops.

Weed lists name crop pests—weeds that cause an economic impact in a county, a state, or a region. These lists are generally available from land grant universities and agricultural extension offices. Herbaria (collections of specimens of plants) and the botanical and agronomic literature that are available at most agricultural universities are repositories of information on the history and occurrence of crops and relatives as weeds. Finally, experts in plant and crop sciences, at universities and in extension education services, will be able to provide information on the weediness of crops and relatives in this and other countries.

Because this kind of information is notoriously scattered and uneven in quality, it is difficult to know how much confidence to place in a "no" answer, that is, a conclusion that a crop is not a weed and has no weedy relatives. A false negative answer is likely where lists are incomplete or inaccurate, or are difficult to find. The USDA's review of a petition for commercialization of transgenic squash reveals the danger of leaping to a "no" conclusion. In this case, the applicant—the Asgrow Seed Company—submitted an analysis of risks to the USDA. Part of Asgrow's analysis quite properly included a consideration of whether transgenic squash would become a weed. Again, quite properly, part of that discussion included a look at whether the crop itself or any close relatives had been reported as weeds in the United States. Asgrow, however, failed to mention that a close relative of squash, Texas gourd, is a serious weed in a number of states[7] (Harrison et al., 1977; Oliver et al., 1983; Wilson, 1993).

The tier 1 analysis separates crops into two risk categories. The lower-risk category contains crops that are not weedy and do not have close weedy relatives in North America. By contrast, the higher-risk group is weedy or has close weedy relatives. The higher-risk crops are subject to a standard set of experiments, while the lower-risk ones undergo an abbreviated set.

Tier 2. Ecological Performance of the Transgenic Crop: Does the Transgenic Crop Outperform the Nontransgenic Crop in Population Replacement Experiments?

This section outlines relatively simple, straightforward experiments to compare the ecological performance of transgenic and nontransgenic crops. To get a picture of performance, the approach asks how well the transgenic population replaces itself over time, compared with the nontransgenic population; that is, is the new genetic type doing better or worse than the parent type?[8] Ecological performance experiments can be run concurrently with the small-scale efficacy tests routinely conducted by companies. By the time a transgenic crop has been shown to be efficacious, the company could have enough data to describe the relative population replacement capacity of the transgenic and nontransgenic crops.

One approach for evaluating ecological performance involves two rather simple measures of how well a population is faring: net replacement rate and seed bank persistence. Net replacement rate is a gauge of a population's growth rate that is determined by counting the number of seeds produced after a specific period of time by a population that developed from a known number of seeds.[9] Seed bank persistence is a measure of how long viable seeds remain in a seed bank; that is, how long seeds are available to germinate and grow into adult plants capable of reproduction (Baskin and Baskin, 1985; Cook, 1980). Seed bank persistence is measured in terms of half-life: how long it takes for the seed bank to lose half its viable seeds.[10]

The emphasis on seed banks reflects the importance of dormancy,[11] seed death, and germination responses to environmental conditions in the persistence of a population (Linder and Schmitt, 1994; Rees and Long, 1992). In fact, the net replacement rate could be underestimated if seeds remained dormant in the seed bank at the time the rate is measured. Seeds might germinate later under more favorable environmental conditions and produce offspring that maintain or increase a population.[12]

Adopting this approach is based on two premises. First, a transgenic plant whose population declines over time relative to a nontransgenic crop is unlikely to become more of a weed problem than a nontransgenic one. Second, because an individual's contribution to population growth depends on the summed effects of each stage of seed production (Manasse and Pinney, 1991), it is unnecessary to evaluate individually a number of different life stages to determine differences among genetic types in population growth rate. The net replacement rate and seed bank half-life measure the culmination of an array of interdependent events: seed germination, seedling survival, vegetative growth, adult survival, reproduction, and others—all influenced by the environment. Of course, measures of some individual life stages would be helpful in understanding the population dynamics of the transgenic crop, but they are unnecessary for following population growth or decline.

The remainder of this section outlines the basic features of one approach to population replacement experiments, explains how the experiments differ for crops designated higher- and lower-risk, and discusses the possible outcomes of the experiments. The appendix describes how such experiments might be conducted. (See also Linder and Schmitt, 1994.)

It is assumed that all tests would be designed with three additional goals in mind: (i) to minimize, to the extent possible, the escape of transgenic plants from field plots or the flow of genes to wild relatives; (ii) to appropriately incorporate relevant biological and ecological factors (for example, selecting experimental sites; sowing, retrieving, evaluating seeds); and iii) to ensure that data can be statistically analyzed.

Two Levels of Experiments: Standard and Abbreviated. Of the two weed strategies described in chapter 3—invasiveness and persistence—this testing scheme emphasizes the former, because invasive weeds are the greater threat to nonfarm ecosystems. This plan encompasses two levels of testing corresponding to the two categories of crops developed in tier 1: one level for lower-risk crops and the other for crops found to be in the higher-risk group. The levels differ in the number of environments where the experiments should be conducted.

In the context of the population replacement experiments proposed in the appendix, "environment" refers to a crop-growing area; that is, a part of the country where the crop is typically or potentially grown (particularly where the transgene extends the crop's range). A crop-growing area is characterized by a particular range of abiotic (climate and soil type) and biotic (plants, animals, microorganisms) factors. Examples of U.S. growing areas are the high plains of Texas, irrigated deserts of Arizona, Salinas Valley of California, the Northern Plains of the Midwest, and the Mississippi Delta.

Within each environment, the tests would be placed in field margins, habitats relatively less disturbed than cultivated fields. Testing in field margins emphasizes the capacity for a crop to invade habitats adjacent to farm fields. This habitat is more challenging than a field site because of the presence of plants that provide greater competition and less human interference that favors the crop over its challengers. Moreover, if a crop does not persist in a field margin, it is not expected to be able to invade more formidable habitats outside the field.

For crops that have a history of weediness in cultivated settings, applicants should also consider conducting replacement experiments in field sites to determine the relative persistence of the transgenic crop; that is, whether it will become a worse weed than the nontransgenic form.

In the approach suggested here, crops determined to be lower risk on the basis of analysis of existing data would be subject to the abbreviated set of experiments: three-year population replacement experiments in three to five growing areas. Higher-risk crops would be tested under the standard regime involving more environments: three-year experiments in the full range of growing environments. The final choices of numbers and kinds of environments for both the abbreviated and standard sets would be made on a case-by-case basis depending on what is known about a particular crop/gene combination and the range over which the crop is expected to be grown.

The abbreviated testing requirement—calling for a reduction in the number of growing area test sites—is based on the premise that potential weediness is low enough in this subset of crops that its expression is

Trade-off: Increased Risk for Increased Information

In some cases, planting transgenic crops next to natural communities may present risks of releasing transgenic plants heretofore not faced in small-scale field trials. Under typical containment strategies for small-scale tests, volunteer transgenic plants are removed, seeds are collected, or crops are bagged or prevented from flowering. These measures, while reducing risk, may preclude the possibility of developing information on the transgenic crops' weediness traits. The additional risks that may accompany the testing near natural areas should be carefully considered before proceeding with the experiments and should be reduced where possible by appropriate containment and emergency procedures. For example, if a transgenic crop or gene appears to be spreading into natural areas, it may be important to remove the offending plants. In some cases, the trade-off is a reasonable one provided the risk is low and relatively controllable.

not likely to be significantly affected by variation in environments. On the other hand, planting higher-risk crops in field margins in a number of environments allows an evaluation of population replacement capacity under a variety of conditions.

At both levels, the selection of crop-growing areas for the tests will depend on how widely the crop is to be grown. A crop restricted to one or two areas of the country need not be subject to testing in as many environments as a crop grown all around the country. For lower-risk crops, it seems reasonable that experiments be conducted in three to five representative environments. By contrast, where weediness is more likely, crops should be evaluated in considerably more environments to determine if risk will differ from growing region to growing region.

Because these tests can be conducted along with the multiyear efficacy tests required of all new crops, the abbreviated approach for the lower-risk group allows a confirmation of the status predicted from tier 1 without a heavy testing burden. The experiments will also provide important empirical data that can ultimately be used to credibly establish categories of plants that may be exempt from tier 2 testing.

An additional modification may be adopted for some lower-risk crops. Companies may omit the seed bank persistence experiment for lower-risk crops where it is known from the published literature that the crops do not produce a seed bank; that is, they show zero percent dormancy. This exception is based on the knowledge that some crops, such as corn, produce no seed bank because dormancy has been bred out of the crop, ensuring even and rapid germination to allow for a uniformly harvestable crop at the end of the season (Hawkes, 1983).

Some people may argue that this scheme is flawed because it does not take into account the nature of the transgene. Some researchers maintain that it is possible to categorize transgenes, a priori, into two broad risk categories: those likely to increase weediness and those unlikely to do so. In the first group are transgenes like herbicide-, insect-, salt-, and disease-tolerance, which may provide an ecological advantage to the crop. Transgenes judged less likely to affect weediness traits are ones, for example, that alter oil, carbohydrate, or protein composition of seeds. The argument follows that transgenes in the second group should be subject to less testing than those in the first.[13]

Neither of these arguments has been substantiated by ecological data on transgenic crops. Moreover, a closer look at some of the transgenes may challenge the theories. For example, the transgenes that alter seeds may, at first glance, appear to confer little ecological advantage to a crop. However, one could reasonably ask whether the changes in seed composition affect seed-related weediness traits, such as dormancy and germination capacity (Linder and Schmitt, 1994).

Once a substantial body of ecological data has been accumulated on transgenic crops, it may be possible to identify categories of "low-risk" transgenes. At that time, risk assessment protocols can be altered to reflect the categorization.

Possible Outcomes. At the end of the three-year tests, companies will have determined net replacement rates and seed-bank half-lives for both transgenic and nontransgenic crops from field margins in several environments. Interpreting the results means first comparing the performances of transgenic and nontransgenic crops in each environment and then synthesizing those results into an overall conclusion.[14]

To obtain the relative performance of the engineered and nonengineered crop, the net replacement rates and seed-bank half-lives are compared in each environment. Each net replacement rate (R) is the ratio of the number of seeds generated by a crop to the number of seeds sown (see appendix). The R values generated will be equal to or greater than zero. A value of zero or a fraction less than one means fewer seeds were produced than sown; that is, these populations are not replacing themselves and will disappear.

A value of one means the crop is just replacing itself and is not increasing. A value greater than one indicates that a population is growing beyond simple replacement. The number indicates how many times the crop has multiplied itself in a three-year period; for example, a value of 3.5 means that the population multiplied three and one-half times. A seed-bank half-life (H) is expressed in terms of months or years and indicates the length of time it takes for half the seeds in the seed bank to either germinate or die. A long half-life means that seeds remain viable in the soil for a considerable period, "awaiting" environmental signals conducive for germination and survival of a new individual. For example, a seed bank with a half-life of three years is more important in the persistence of a population than a seed bank with a two-month half-life.

A simple comparison of replacement rates and half-lives of the two plant types tells which is performing better in a particular environment. Then, the task becomes one of interpreting the results across the spectrum of environments. Conclusions about the relative overall performance of the transgenic and nontransgenic crop should fall into two categories: lower-risk, where the transgenic performs no better than the nontransgenic; and higher-risk, where the transgenic outperforms the nontransgenic. Where relative performances are consistent across environments, categorization is straightforward. More problematic are situations in which the transgenic may outperform the nontransgenic in some environments but not others.

Lower-Risk Outcome. An engineered crop will achieve a lower-risk designation when it performs no better than the nontransgenic in all environments.[15] A crop that is not replacing itself[16] also qualifies for a lower-risk designation. If the net replacement rate for the transgenic is consistently less than one, the crop may be considered lower risk, regardless of the replacement rate of the nontransgenic, because the transgenic is not replacing itself and will disappear over time. Transgenic crops placed in the lower-risk category require no further evaluation of weediness potential.

The only caveat to this statement is whether the transgenic seed bank might compensate for the crop's decline and alter its performance relative to the nontransgenic one. If the seed-bank half-life of the transgenic is less than or equal to the nontransgenic, the transgenic can remain in the lower-risk category. If however, the half-life of the transgenic bank is greater than that of the nontransgenic, then additional consideration is warranted. If the seed bank appears to be able to compensate for the decline, then the crop should go to the higher-risk category.

Higher-Risk Outcome. A number of different results of population replacement experiments may cause transgenic crops to be assigned to the higher-risk category. The designation is given to a transgenic crop that consistently replaces itself at a rate greater than the nontransgenic.[17] With this replacement capacity, the relative seed bank persistence is unimportant.

Even where performances are not uniform across all test sites, the prudent approach would be to place any transgenic crop in the higher-risk category if it outperforms the nontransgenic in any environment. Admittedly, this is a conservative position; but it is reasonable, given the lack of experience and ecological data on engineered crops in the environment. It may be possible, once sufficient ecological data are generated, to loosen the stringency of this approach.

As a general principle, companies should be discouraged from commercializing crops where the transgenic crop consistently outperforms the nontransgenic, as enhanced replacement capacity sets the stage for adverse impacts. Critics may argue, at first glance, that these tests would prevent the commercialization of many transgenic crops because they have been developed in many cases specifically to outperform the nontransgenic under field conditions. They would argue, correctly, for example, that transgenic herbicide-tolerant rapeseed should outperform nontransgenic rapeseed under field conditions where herbicides were applied. The approach proposed here, however, is to compare crops under nonfield conditions, that is, in field margins, where herbicides would not be applied. It should be noted that under these conditions, many crops, engineered or not, are not likely to replace themselves over a three-year period. The capacity for the transgenic crop to outperform the nontransgenic is unimportant if the transgenic is not replacing itself. But in cases where the transgenic can replace itself, this experiment would serve to detect unexpected ecological advantages conferred on the transgenic crop.

Some companies may believe that their transgenic crop, particularly one that outperforms the nontransgenic in only one or two sites, or under particular conditions, is unlikely to become a weed. In this case, companies may decide to proceed to a tier 3 analysis; that is, the transgenic crop would be subject to additional experimentation to determine to what extent and under what circumstances the crop poses weediness problems.

In some cases, companies may initiate additional experiments before the end of the three-year tests. For example, if data from the first and second year show that the transgenics are outperforming the nontransgenics in only one or two sites, companies may want to gain an advantage by beginning an early experimental evaluation of the relationship between enhanced ecological performance and weediness.

In sum, the tier 2 analysis relies on relatively simple experiments—which can be conducted along with efficacy tests—to evaluate the performance of transgenic crops relative to nontransgenic ones. This measurement uses empirical data to separate transgenic crops into lower- and higher-risk categories, that is, crops that are not likely to become weeds and those that may have a higher potential for weediness. For crops designated lower risk, the analysis of weediness ends. Developers of crops likely to become weeds should either reconsider commercialization or subject the crop to tier 3 testing.

Tier 3. Weediness of the Transgenic Crop: Is Weediness Increased in Transgenic Crops Exhibiting Enhanced Ecological Performance?

The tier 3 analysis is designed to determine whether superior ecological performance translates into enhanced weediness. Because a crop with increased performance will not necessarily become a weed, companies should have the option of establishing the safety of the crop experimentally. To do so would probably require multiyear, confined small-scale field tests in a number of environments.

The specific design of experiments will depend, to a great extent, on the data obtained from tier 2 experiments and what is already known about the particular transgene and crop under consideration. Considerable consultation among companies, regulators, agronomists, ecologists, and statisticians will be required to formulate tier 3 experiments.

One source of guidance is the Planned Release of Selected and Manipulated Organisms (PROSAMO) initiative completed in Great Britain (Crawley et al., 1993). Within the PROSAMO program, Crawley and his colleagues at Imperial College in England developed experimental data that compared transgenic herbicide-tolerant rapeseed (canola) to the nonengineered form. Specifically, they conducted a three-year, three-environment, twelve-habitat study comparing two transgenic lines and a nonengineered variety. The major objective was to evaluate the transgenic crop's potential for weediness in a number of habitats across the range of environmental conditions where the crop is grown or could potentially be grown (Crawley, 1990b, 1992; Crawley et al., 1993; Rees et al., 1991).

The study measured a number of phenomena including seed dispersal, seed mortality, seedling growth, and herbivore and pathogen impact. After evaluating these parameters in transgenic and control plots, the scientists found no indication that genetic engineering for herbicide tolerance increased the invasiveness of canola; rather they found in some cases a significant difference—such as seed survival in soil—where transgenics performed less well than nontransgenics (Crawley et al., 1993).

If the tier 3 experiments reveal why the transgenic crop outperforms the nontransgenic and why it will nonetheless not become a weed, then the crop can be considered lower risk for commercialization. If, however, the tests show that enhanced performance does, in fact, translate into enhanced weediness, then the developer should reconsider commercialization of that crop.

The tier 3 analysis allows developers an opportunity to demonstrate that transgenic crops that outperform nontransgenics in the ecological performance tests do not pose risks as weeds under conditions of commercial use.

Assessing the Potential for Transgene Flow to Produce Weeds

Like the previous assessment strategy, assessing the second major environmental risk of transgenic crops—gene flow and its consequences—involves three tiers of analysis. The first tier, using existing information for the most part, analyzes the likelihood that transgenic hybrids will form between the transgenic crop and wild/weedy relatives. The second and third tiers, which are the same as the second and third tiers in the assessment of weediness potential in transgenic crops, rely on experimental data on transgenic wild/weedy plants compared with the nontransformed counterparts.

The first section below outlines the first-tier assessment—the analysis, sources of existing information, and experiments needed to determine whether transgenic hybrids are formed. If the tier 1 assessment indicates no hybridization, the transgenic crop is considered to be low risk, and the gene flow assessment ends here.

If viable, fertile transgenic hybrids are formed, the assessment moves to the second tier, which requires experiments to determine the ecological performance of transgenic and nontransgenic wild/weedy plants. If the transgenic wild/weedy plants perform no better than their nontransgenic counterparts, the gene flow assessment ends here. Where the transgenic wild/weedy plants outperform the nontransgenic ones, companies should be discouraged from commercializing the crop from which the genes came.

If companies believe that the enhanced performance of the wild/weedy plants will not translate into weediness, they may choose to conduct the third-tier analysis. Figure 4.2 outlines the analysis.

Tier 1. Gene Flow: Do Viable, Fertile Hybrids Form between the Crop and Wild/Weedy Relatives?

Like the tier 1 assessment in the weediness section, the analysis relies primarily on existing information. Specifically, through a series of yes/no questions, it assesses the likelihood that viable, fertile transgenic hybrids will form between a crop and wild/weedy relatives. The meaning of each question and the data needed to answer it are discussed below. Except for the last question, which requires experimentation, information generally can be sought in the botanical and agronomic literature. In some cases, the genetics literature may contain reliable information on hybridizations between crops and relatives. When crossing experiments conclusively show that crop/relative hybrids do not form, are inviable, or are sterile, it may be possible to eliminate the gene flow analysis for that particular crop/relative combination.

Taken together, the questions address the factors that will determine whether a transgene is likely to be passed to a wild/weedy plant-crop hybrid. These include considerations of crop sexuality, cross-compatible relatives in the United States, and flowering and pollination characteristics. Although each issue is approached with a simple yes/no question, a full analysis may be more complex. The evidence supporting a "no" answer must be especially persuasive, as it ends the analysis. Where there is insufficient information to make a determination, the analysis should proceed to the next step.

As with the tier 1 analysis in the weediness section, regulatory agencies would ultimately determine the potential for each U.S. crop to form hybrids with wild/weedy relatives. Once these analyses are available, companies could eliminate the first-tier assessment.[18]

Tier 1: Gene flow
Do viable, fertile hybrids form between the crop and wild/weedy relatives?

a. Is the crop sexual?

Yes/Insufficient information	No
Go to b	Lower risk
	End of analysis

b. Does the crop have sexually compatible wild/weedy relatives in the United States?

Yes/Insufficient information	No
Go to c	Lower risk
	End of analysis

c. Do the crop-relatives breeding systems permit gene flow in and out?

Yes/Insufficient information	No
Go to d	Lower risk
	End of analysis

d. Does the flowering phenology of the crop and wild/weedy relatives overlap, or nearly so?

Yes/Insufficient information	No
Go to e	Lower risk
	End of analysis

e. Do crop and wild/weedy relatives share the same means of pollination?

Yes/Insufficient information	No
Go to f	Lower risk
	End of analysis

f. Do crop and wild/weedy relatives naturally cross-pollinate, fertilize, and set viable, fertile seeds under field conditions?

Yes	No
Go to tier 2	Lower risk
	End of analysis

Tier 2: Ecological performance of transgenic wild/weedy plants
Do transgenic wild/weedy populations outperform the nontransgenic wild/weedy plants in population replacement experiments?

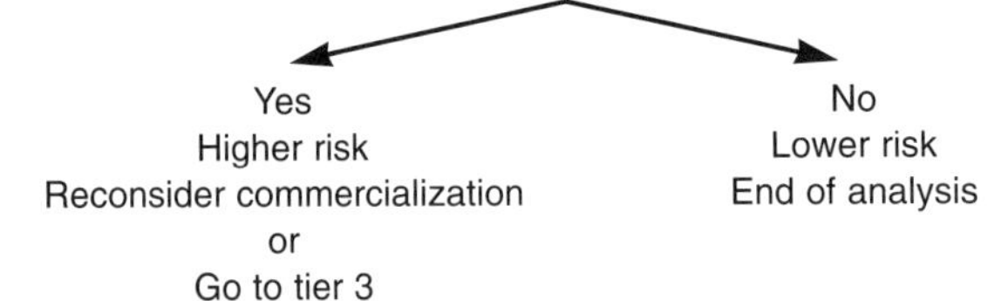

Tier 3: Weediness of transgenic wild/weedy plants
Is weediness increased in transgenic wild/weedy plants exhibiting enhanced ecological performance?

Yes	No
Higher risk	Lower risk
Reconsider commercialization	End of analysis

Figure 4.2
Experimental assessment of the potential for transgene flow to produce weeds in noncrop populations

Is the Crop Sexual? This question addresses whether or not a crop reproduces sexually. Of particular interest is whether the crop produces viable pollen, that is, pollen that can germinate and produce a pollen tube. If a crop is known not to sexually reproduce or produce viable pollen, the analysis of gene flow ends here.

Plants that are not sexual reproduce exclusively by asexual or vegetative means. In reality, however, few crops are completely nonsexual. In the United States, some citrus, forage, and turf grass cultivars may qualify for a "no" answer to this question. The botanical and agronomic literatures are likely to provide supporting data for the crop's lack of sexual reproduction or viable pollen.

Does the Crop Have Sexually Compatible Wild/Weedy Relatives in the United States? For U.S. crops that have no sexually compatible relatives in the United States, the risk assessment ends here. Many crops, however, have relatives somewhere in the country. For some of these, the likelihood of outcrossing could nevertheless be virtually nil if the transgenic crop and relatives are shown to be sexually incompatible.

In some instances, the existence of crop relatives in Hawaii or other U.S. islands may warrant special consideration. A crop that has relatives only in Hawaii or the other islands and nowhere in the continental United States may be placed in a low-risk group provided there is assurance that the crop will never be grown on the islands.

The key problem here is substantiating claims that no interfertile relatives reside in the United States. Three major sources of information on possible relatives are the agronomic and botanical literature, crop breeders, and taxonomists. For some crops, like soybeans, claims can be supported fairly easily. For other crops, however, the information on crop-wild/weedy plant compatibility is scattered in the literature and in county and state departments of agriculture. In general, reliable distribution maps are lacking for U.S. crops and their weedy relatives (see, for example, Boyce Thompson Institute, 1987; Wilson, 1990). In addition to the difficulty of assembling information from diverse sources, there are, in many cases, serious gaps in the knowledge of interfertility.[19]

In many cases, determining interfertile relatives, or lack thereof, for many crops could require the generation of significant amounts of information, such as the distribution of wild relatives and the sexual compatibility of each relative with the crop. Much of this information would have to be produced from new surveys and greenhouse and field studies because existing information is scarce for most crops (Boyce Thompson Institute, 1987; Mack, 1988). In some cases, the needed field work may be so extensive that companies might prefer to assume a "yes" answer and proceed with the risk assessment.

Some may question why the search for relatives extends to the entire United States and is not limited to areas where the crop will be grown commercially. A broad search is admittedly a conservative approach, but one that provides a margin of safety, particularly in light of the fact that commercialization means that growers and home gardeners will be free to plant anywhere in the country.

One question that arises is how to define the subset of plants that should be examined for their interfertility with a crop. For regulatory purposes, generic classification—that is, wild plants placed in the same taxonomic genus as the crop—rather than species, should be the criterion used in subjecting plants to investigation for interfertility.[20] Although it is true that, in general, sexually interbreeding relatives share species designation, there are many instances of interbreeding among different species belonging to the same genus or even among different species in closely related genera. For example, *Cucurbita adreana*, a toxic and weedy species, crosses with some edible squash species (*Cucurbita maxima*) (Nee, 1990).

Furthermore, wild relatives may be named as a species different from the crop on the basis of incomplete data or on judgments of taxonomists who may take many factors into consideration in addition to interfertility. Thus, the relatives may, in fact, be sexually compatible with the crop, yet be placed in another species or even genus. Using genus rather than species to define relatedness provides a margin of safety.[21]

Information from several sources may be used to determine whether a crop has relatives in the same genus in the United States:

i. Botanical and agronomic literature; up-to-date, complete, reliable floristic surveys; and/or herbaria;

ii. New surveys; and

iii. Experts, including crop breeders, weed scientists, economic plant botanists, and plant taxonomists.

Do the Crop-Relatives Breeding Systems Permit Gene Flow In and Out? This question addresses the extent to which crops and relatives "accept" pollen from genetically different plants, that is, the extent to which they are self-pollinators (inbred) or cross-pollinators (outcrossers). Self-pollination refers to pollination where the pollen comes from the same plant that produces the egg or from other plants with the same genetic makeup. Cross-pollination means that pollen and egg come from plants that are genetically different. The genetic differences may involve one or many characteristics. The parents may be the same variety, the same species, or different species and even genera. Cross-pollinating plants permit gene flow *in and out.*

Different plant species are located at different places on a continuum between "exclusively" self-pollinating to "exclusively" outcrossing. Rapeseed, or canola, is a partially cross-breeding crop whose outcrossing rate varies widely depending on cultivars, distances between plants, and environmental conditions. There are reports, for example, of outcrossing rates of 5 percent to 15 percent in winter oilseed rape under field conditions in Germany and 22 percent to 23 percent in spring canola in Canada (Scheffler et al., 1993 and references therein).

Even the extremes are not absolute because "exclusively" self-pollinators outcross, though rarely, and "exclusively" cross-pollinators occasionally self-pollinate. In the United States, for example, wild oats (*Avena fatua*) are largely inbred but occasionally outcross (K. Keeler, personal communication, 1992). Tomatoes are highly, but not absolutely, self-pollinating (Simmonds, 1976). Although they outcross considerably in some parts of Central and South America and other subtropical areas, they are almost completely self-pollinating elsewhere (for example, North America) (Simmonds, 1976 and references therein). Even largely sterile species like the sweet potato probably outcross occasionally (K. Keeler, personal communication, 1992).

Even very low flow rates and rare events become important when a crop is planted in the hundreds, thousands, and millions of acres. A hypothetical example illustrates the problem. An outcrossed seed occurs at a 0.01 percent rate in a largely inbred crop. The crop contains an average of 100 flowers per plant with 100,000 plants per acre. If that crop is planted in 1 million acres, then 1 billion outcrossed seeds are produced annually (K. Keeler, personal communication, 1992).

The breeding system of most crops grown in the United States is known and can be determined from the agronomic literature or crop breeders. Even in the United States, however, it may be more difficult to determine the breeding system of wild/weedy relatives, simply because little ecological research has been devoted to many of these plants (Boyce Thompson Institute, 1987). Where information is lacking, greenhouse studies may be required to determine their inbred/outcross status.

Does the Flowering Phenology of the Crop and Wild/Weedy Relatives Overlap, or Nearly So? Flowering phenology refers to the timing and amount of pollen production. Fertilization and production of wild/weedy plant-crop hybrid seed are likely when a crop produces pollen at the same time that the female part of the flower of a nearby compatible relative is receptive to pollen. On the other hand, crop and compatible weedy relatives that flower weeks apart in the same season or in different seasons are not likely to produce crop-weed hybrid seed. Others fall between the two extremes of the spectrum. For example, Langevin et al. (1990) suggest that differences in hybridization rates among cultivated rice and weedy red rice (both *Oryza sativa*) and their hybrids may be due to variation in the degree of overlap of flowering times among them.

In rare cases, crops and wild/weedy plants may flower on the same days but at different times of day—for example, the crop may produce pollen in early morning while the female flower parts of relatives are open and receptive in late afternoon. Despite the difference in timing and the fact that pollen is generally short-lived, these diurnal differences are not enough to guarantee that no cross-pollination and fertilization would occur.

Practically speaking, since flowering is fraught with considerable variability, it may be difficult to substantiate a claim that the phenology of a crop and its relatives does not overlap. The agronomic and botanical literature, agronomists and botanists, and direct observations may provide information to address this phenomenon.

Do Crop and Wild/Weedy Relatives Share the Same Means of Pollination? Pollen is disseminated primarily by wind and insects. A crop and compatible weed that do not share the same means of pollen dispersal—for example, the crop is pollinated by honey bees and the noncrop relatives by butterflies—are less likely to produce hybrids than a crop and weed with a common dispersal agent.

For the most part, however, pollination biology is not this simple. Very few plants are pollinated by only one kind of insect. Generally, several species of insects pollinate a plant species (P. Regal, personal communication, 1993). Moreover, wind-transported pollen may land on a normally insect-pollinated flower. And insects that gather nectar from "insect-pollinated" flowers may go to certain "wind-pollinated" plants and in theory could be pollinating.

Although data on crop pollination in the United States may be available in the agronomic literature, this information may be generalized from studies of major pollinators and fail to take into account the contributions of rare pollinators. Pollinators for wild/weedy plants in the United States are even less likely to be known. In the absence of reliable existing data, pollination could be determined through careful field observations. Rather than gather these data, companies may wish to go directly to the next question.

Do Crop and Wild/Weedy Relatives Naturally Cross-Pollinate, Fertilize, and Set Viable, Fertile Seeds under Field Conditions? Thus far, the line of questioning focuses on phenomena that may have been described in the agronomic or botanical literature or that can be tested under greenhouse conditions or observed in the field. This question is the first to address the likelihood that the crop and sexually compatible wild/weedy relatives produce viable, fertile hybrid seed under field conditions. For the plants that have not fallen out of the analysis as a result of answers to earlier questions, answering this question requires field experimentation.

To answer this question, the crop and wild/weedy relatives are grown next to each other in a field at a number of locations, seeds are harvested, and a determination is made whether wild/weedy plant-crop hybrids have formed. The experiment should be done in at least two different years. The design of this experiment will depend upon existing information in three areas: sexually compatible relatives, natural crossing rates, and hybrid viability.

Sexually Compatible Relatives. The first step in designing the experiment for U.S. crops is to determine the sexually compatible relatives classified in the same genus. Two routes are possible. First, if reliable information is lacking in the literature, the compatibility of relatives is evaluated in greenhouse experiments where crops and relatives are crossed. With data identifying cross-compatible relatives in the same genus, the next step is to determine if any of these cross-compatible relatives exist anywhere in the United States. The last step depends on the existence of up-to-date, reliable floristic analyses, local weed surveys, or expert knowledge. In the absence of reliable information, new plant surveys will be needed.

The other route is to take the list of potentially cross-compatible relatives and, using existing information or new surveys, determine whether they reside in the United States. The next step then is to determine whether the potentially compatible relatives are in fact cross-compatible. Unless there is reliable information in the literature, data are best obtained through experimental crosses in the greenhouse. Where extraordinary means, such as embryo rescue, are required to obtain crosses between a crop and its relatives, the two plants would be considered incompatible.

In some cases, the genetics literature may contain reliable studies with sufficient data to establish lack of hybridization under field conditions between crops and some relatives. Because of the special considerations given to protecting endangered species in this country, the endangered species status of any potentially sexually compatible wild/weedy species should be determined.

Crossing Rates under Field Conditions. Crop breeders generally know a lot about gene flow *into and within* a crop. Many older studies of gene exchange were performed to determine the degree to which coexisting pure lines of seed crops would contaminate each other. As a result of this fundamental work, plant breeders and seed producers know, with or without information on interbreeding with wild relatives, how to produce seed with a low level (but not zero) of genetic contamination from relatives or other crop cultivars—by controlling the pollen flowing into the crop. Generally, this is accomplished by isolating breeding plots or seed production fields sufficiently far from other cultivars or relatives to ensure that no, or only very low levels of, pollen from other sources contaminate the "pure" seed. The particular isolation distances vary depending on the crop and what is known of pollen dispersal agents.

Much less is known about the rate of gene flow, under field conditions, to relatives from the crop. Some recent work is beginning to address these gaps and provides models for additional research. For example, Wilson and co-workers planted seedlings of Texas gourd (*Cucurbita pepo* ssp. *texana*) within pollinating distance of cultivated squashes (*C. pepo*). After allowing flowering, pollination (by insects), and fruit maturation, an analysis of the seeds found that 5 percent of the progeny (seeds or seedlings) contained genes from both wild and cultivated parents—results that established definitively that gene flow had occurred between wild and cultivated plants (Kirkpatrick and Wilson, 1988).

In related research, Langevin et al. (1990) documented hybrids between cultivated rice and weedy red rice (both *Oryza sativa*) varying from 1 percent to 52 percent depending on the cultivated rice cultivar. They found moderate levels of hybridization between weedy rice and five of the six rice cultivars—ranging from one to less than 8 percent. For the sixth cultivar, they found a much higher rate—over 52 percent. They speculate that this higher rate is due to a longer overlap of flowering times of the interbreeding plants (Langevin et al., 1990). Klinger et al. (1992) analyzed gene flow from cultivated radish into wild radish (both *Raphanus sativus*). Crop-to-weed gene flow was extensive for weed plots planted at the edge of the crop field and then diminished considerably in weed plots at greater distances—200 to 1,000 meters—from the crop. However, some gene flow was detected in hybrids at 1,000 meters. They also noted that the rate of hybridization varied significantly depending on the size of the recipient population; that is, small populations had greater hybridization rates than larger populations at one meter distance. At long distances (400 meters), however, larger populations showed greater rates than smaller ones (Klinger et al., 1992).

Even in the largely self-pollinating crop-weed system of cultivated foxtail millet (*Setaria italica*) and wild green foxtail (*S. viridis*), hybrids occurred at rates between 0.002 percent and 0.6 percent (Till-Bottraud et al., 1992). Kareiva and co-workers, based on field experiments with wild mustard (*Brassica campestris*), are developing models that can, among other things, predict gene flow in insect-pollinated plants (Kareiva et al., 1991; Manasse, 1992).

Hybrid Viability. Hybrid viability is known for some wild/weedy plant-crop combinations and can range from complete inviability to full viability.

In some cases, levels of viability may be unknown. But the foregoing field experiments will already have shown whether or not hybrid seed could be produced under agricultural conditions; these seeds could then be tested for viability.

More problematic is the result of a single field test where no hybrid seeds are formed. Consideration should be given to additional field experiments under different environmental conditions that may allow pollination and fertilization to occur.

The tier 1 analysis determines whether transgenic hybrids will form between transgenic crops and their wild/weedy relatives. Where hybrids are not formed, the transgenic crop is deemed to pose low risk in terms of gene flow and no further tests are required. Where hybrids are formed, the analysis moves to tier 2. Once gene transfer occurs, the assessment of potential adverse impacts is the same as for the transgenic crop itself.

The next part of the analysis focuses on the wild/weedy plant x crop hybrids containing the transgene and wild/weedy plants into which the transgene has introgressed.[22]

Tier 2. Ecological Performance of Transgenic Wild/Weedy Plants: Do Transgenic Wild/Weedy Plants Outperform Nontransgenic Wild/ Weedy Plants in Population Replacement Experiments?

Like the counterpart in the assessment of transgenic crop weediness, the tier 2 analysis of gene flow calls for a measure of the relative ecological performance of transgenics, in this case transgenic wild/weedy plants.[23] Because so little is known about wild/weedy plants into which crop genes have introgressed, it is not possible to carve out subsets that qualify for the abbreviated set of replacement experiments. Thus, ecological performance tests would be conducted in environments that represent a full range of the crop growing areas where compatible relatives reside (Linder and Schmitt, 1994). (See the discussion of tier 2 experimentation above for additional details on this level of analysis.)

As proposed above, ecological performance is measured by evaluating net replacement rate and seed-bank half-life.[24] Unlike the earlier assessment, however, an additional consideration is needed here early in the experimental design. The question is: which generations of transgenic wild/weedy plants should be tested? F_1 hybrids?[25] how many backcrossed generations?[26] Put another way, the question is: what is the genetic background in which the transgene is to be tested? In this case, genetic background refers to the extent to which the offspring containing the transgene also contain other crop and wild plant genes (see "Becoming Wild").

Becoming Wild

The F_1 hybrid typically contains 50 percent crop genes and 50 percent wild plant genes. (A wild plant may, in fact, carry some crop genes, depending on the extent to which the wild population has been subject to gene flow from previous crops.)

A first generation backcross (the hybrid x wild parent) will have an average of 75 percent of its genetic material from the wild/weedy plants. Subsequent backcrossed generations will be successively wilder; for example, the second generation averages 87.5 percent wild genetic material, the third 93.75 percent. Only a small number of generations is needed before the genetic makeup of the backcrossed offspring is nearly all wild. If the transgene is retained during backcrossing, it eventually will be acting in a genetic background that is largely wild/weedy.

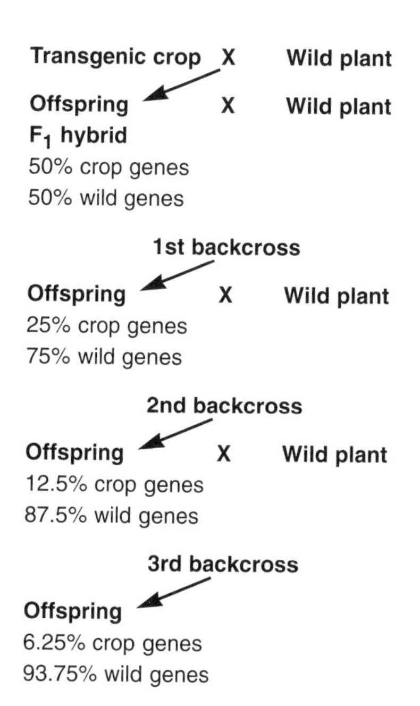

To determine which generations to assess, one considers the particular situation with a specific crop and its relatives. The situation might be one where pollen flow from the crop to relatives is a common annual event. Or it might be the other extreme, where pollen flow is a rare event.

The first case is illustrated by a crop like canola, which is grown to flowering, produces large amounts of pollen, hybridizes with wild/weedy relatives, and is grown in the same area year after year. This means a large annual flow of pollen into wild/weedy populations. In this case, the wild populations may have a fairly large complement of crop genes in their gene pool. With crops like these, the population replacement experiments would be conducted with offspring containing a fair amount of crop genes, that is, one or two backcrossed generations.

The second case, where gene flow is rare, is typified by a crop like broccoli, which is not generally grown to flowering and therefore produces little pollen. Nearby relatives of these crops would probably not have a large complement of crop genes. A testing scheme for these crops would focus on the third or fourth backcrossed generations.

Before proceeding to field experiments, greenhouse experiments should be undertaken to determine whether it is possible to produce viable and fertile backcrossed or F_2 plants. In some cases, the F_1 may be viable and fertile but the F_2 is not (Levin, 1978). If reproduction cannot proceed beyond the F_2, then the analysis of further transgene spread ends here.

It should be noted that even if the F_1 hybrids are less successful in producing offspring than the weeds, further rounds of interbreeding can change the genetic background in which the transgene acts and thereby alter its fitness effect on the plants. Hence, later generations of plants carrying the transgene may be more or less successful than the initial set of hybrids.

The analysis of gene flow ends here where transgene-containing wild/weedy plants perform no better than their nontransformed counterparts.[27] Where they outperform the nontransgenic ones, the company should be discouraged from commercializing the transgenic crop because enhanced performance may well translate into weediness.

However, additional consideration may be justified in some situations. Some companies may believe that transgenic wild/weedy populations, particularly ones that outperform nontransgenic ones in only one or two sites, are unlikely to become weeds. In this case, the analysis would proceed to tier 3, where the transgenics would be subject to additional experimentation.

The tier 2 analysis experimentally examines the relative ecological performance of transgenic wild/weedy plants. Those that do not outperform their nontransgenic counterparts are deemed low risk and the analysis of gene flow ends. As a general rule, developers should reconsider commercialization of transgenic crops if gene flow from them produces transgenic wild/weedy plants that outperform the nontransgenic wild types. Some situations may be exceptions to the general principle, and companies may proceed to a tier 3 analysis.

Tier 3. Weediness of Transgenic Wild/Weedy plants: Is Weediness Increased in Transgenic Wild/Weedy Plants Exhibiting Enhanced Ecological Performance?

For the few assessments that get this far, tier 3 gives companies an opportunity to examine the hypothesis that enhanced ecological performance does not translate into enhanced weediness. The section above on weediness potential of transgenic crops contains additional discussion of the tier 3 analysis.

Assessing Virus Resistance and Other Aspects of Weediness

The challenge in evaluating virus resistance and other aspects of weediness is immense. As discussed in chapter 3, these risks are difficult to define, much less reliably assess. Research to develop methods for assessing cascading and cumulative effects in the environment has barely begun for other well-known potential disrupters—let alone for genetically engineered products. This section briefly considers the current capability to assess four risks: potential impacts of virus-resistant plants; cumulative and cascading effects; nontarget impacts; and secondary effects on agricultural ecosystems.

Once a transgene is in a plant, the assessment of its potential adverse impacts is the same whether the gene was transferred to a crop by gene-splicing in the laboratory or to a wild relative by pollen flow. Therefore, this section does not differentiate between transgenic crops and transgenic wild/weedy plants.

Risks of Transgenic Virus-Resistant Plants

From the discussion of virus-resistant plants in chapter 3, it appears that plant virologists should be able to design protocols to assess risks potentially associated with some of these transgenic plants. Based on current knowledge, it may be possible, to identify low-risk crop/virus gene combinations or to engineer particular virus-resistant crops so that they present little risk (Palukaitis, 1991; Tepfer, 1993). The number of virus-resistant crops heading toward the marketplace makes it imperative that risk assessment schemes be developed and used before huge acreages of the crops are planted.[28]

Cumulative and Cascading Effects

Evaluating cumulative and cascading impacts of organisms containing new genes is the major challenge confronting those who wish to evaluate fully the risks of this technology. Although regulators will be presented with these issues on a product-by-product basis, the environment will experience them together in various combinations over time. As is often the case, the combined impacts that cause the greatest concern are the most difficult to assess. For the most part, the risks of ecological disruption must simply be acknowledged. They cannot be predicted or controlled.

Recognition of their existence does have a place in the regulatory response to a new technology. First, the awareness of long-term threats to ecological stability should lead governments to err generally on the side of caution in regulatory judgments. Second, it places an increased burden on those promoting a technology to demonstrate its benefits.

Impacts of Pesticidal and Pharmaceutical Transgenes on Nontarget Organisms

It may be possible to begin formulating some kind of rough assessment scheme to evaluate nontarget effects of pesticide- and pharmaceutical-containing transgenic plants. The evaluation, in part, would rely on approaches developed to assess nontarget effects of microbial pesticides (see, for example, Flexner et al., 1986 and Miller, 1990).

An analysis of the potential for a pesticidal or pharmaceutical transgenic crop or wild/weedy plant to affect nontarget organisms focuses, for the most part, on the product of the transgene. Table 4.1 lists questions that may guide the risk assessment of nontarget effects.

Table 4.1
Analysis of effects of transgenic pesticidal and pharmaceutical plants

1. What organisms are the targets of the gene product?
2. What is the mode of action of the transgene product?
3. How much of the gene product is required to have its intended effect?
4. How long and where does the gene product persist in an active form?
5. To what does the gene product degrade?
6. Is expression/production of the gene product affected by the environmental/agricultural context?
7. Does the gene product have an effect like any other known gene product?
8. What is the fate of the transgene?

First, the targets of the engineering, that is, the organisms to be killed or otherwise affected by the pesticidal/pharmaceutical gene products, must be identified. Then, the analysis looks to relatives of the targets to narrow the universe of potential nontarget organisms. The assessment then considers which organisms could be exposed to the product. For example, if the target of the gene product is a pest beetle, then the evaluation considers related nonpest beetles as potential nontarget insects that will be adversely affected by the gene product. Similarly, if a soil fungus is the target, then the analysis focuses on other soil fungi as potential recipients of gene product toxicity. If a crop is engineered to produce human growth hormone, the assessment might focus on other mammals, such as deer or elk, that might consume the crop.

In some cases, a gene product may have effects that require additional nontarget testing. For example, a chitinase (an enzyme that degrades chitin, a component of the cell wall of fungi) may be targeted at soil fungi. Yet, the enzyme may be active on the chitin that makes up the outer skeleton of insects living in the soil.

Questions 2 through 5 of table 4.1 focus on the activity and fate of the pesticidal/pharmaceutical gene product: the amount needed for biological activity, the mode of action of the product, the time and place where it is active, and its degradation products. Answers to these questions determine to a great extent the likelihood that nontarget organisms would be affected. If the pesticide or pharmaceutical is produced in

Can Postrelease Monitoring Reduce Prerelease Data Requirements?

The participants at the experts' workshop discussed the possibility of substituting postrelease monitoring for prerelease risk assessment as a way of controlling the risks of transgenic plants. Some scientists have argued that if an effective monitoring and response system were in place, then prerelease testing to predict and avoid risks could be reduced. Ideally, monitoring could function as an early warning system. In the case of transgenic crops, postrelease monitoring would follow transgenic crops and transgenes looking for unexpected, unwanted effects. If such effects were detected in time, the plants and/or the transgenes might be controlled, perhaps even eradicated.

An effective monitoring strategy would (i) tag and track both the transgenic plant and transgenes; (ii) identify and measure appropriate endpoints for following potential ecological impacts, such as weediness and nontarget, cumulative, and cascading effects; and (iii) eradicate troublesome plants in agricultural and nonagricultural ecosystems.

Workshop participants noted two major problems with substituting monitoring for prerelease risk assessment. First, monitoring is difficult, complex, and expensive when dealing with thousands of acres of transgenic crops. The biggest obstacle is identifying measurable endpoints for nontarget, cumulative, and cascading effects.

Second, remediation to the point of eradication is virtually impossible once transgenic crops are used on thousands and millions of acres. Only in small, localized areas might it be possible to eradicate unwanted plants. In general, invasive plants have proven difficult and expensive to control (Keeler and Turner, 1991), much less to eradicate.

Given these currently insurmountable problems, the workshop participants agreed that postrelease monitoring is not a practical means for controlling risks of commercial-scale uses of transgenic crops. Thorough prerelease testing is both safer and more cost effective than monitoring as a way of controlling the risks of transgenic plants.

There was, however, interest in the possibility of marking transgenes as a way of determining whether plants known to be problems in fact contain transgenes. For example, the government could require transgenes to carry a unique molecular sequence that would serve to identify the gene either in its original host or after transfer to a wild/weedy relative. The tags could facilitate tracking the plants in nature. A central repository of transgene identity tags could also enable society to hold accountable the company that produced the transgenic plant in the first place.

amounts sufficient to exert its desired physiological function, and if sensitive nontarget organisms are within the zone of activity of the product, then the nontarget organisms are likely to be affected.

Question 6 probes the conditions under which the gene product is synthesized and active, that is, whether production and expression are dependent on particular environmental conditions. For example, some products may be active only under certain ecological conditions, such as during a dry or wet year, or in the spring or fall. Knowledge of these conditions helps to determine the likelihood and timing of nontarget effects.

Question 7 seeks information from analogous situations, that is, experience with biological pesticides or human drugs on nontarget organisms. Finally, question 8 addresses the fate of the transgene itself, that is, whether the DNA is totally degraded along with plant debris or could transform indigenous bacteria.

Secondary Effects on Agricultural Ecosystems

Experience with integrated pest management (IPM) may inform an analysis of the secondary effects of transgenic crops on the management of agricultural ecosystems. IPM practitioners have found that changing one component of a pest control system may alter other parts. For example, cover crops are plants used prior to a crop to avoid erosion and weed problems. Often, the use of a particular crop has been found to decrease insect or disease problems in one situation and increase the same or other pest problems in other situations.

Current research in IPM, biological control, and other related areas is attempting to take into account the effects of pest-control measures, not on just two or three organisms, but on a larger set of plants, animals, and microorganisms that make up the affected ecosystem (Duffey and Bloem, 1986; Kennedy, 1986; Kogan, 1986).

Summary

Chapter 4 proposes a novel approach for assessing two aspects of the ecological risks of transgenic crops identified in chapter 3—weediness and gene flow. Figure 4.3 summarizes the scheme.

Risk assessment questions		Sources of data/information
Assessing the potential for transgenic crops to become weeds **Tier 1:** **Weediness of the parent crop**	**Assessing the potential for transgene flow to produce weeds** **Tier 1:** **Gene flow to wild/weedy relatives**	
Is the parent crop weedy or does the parent crop have close weedy relatives in North America? Yes/Insufficient information → Higher risk, Go to tier 2, Standard set No → Lower risk, Go to tier 2, Abbreviated set	**Do viable, fertile hybrids form between the crop and wild/weedy relatives?** Yes → Higher risk, Go to tier 2, Standard set No → Lower risk, End of analysis	Existing information in the botanical and agronomic literature Field experiments

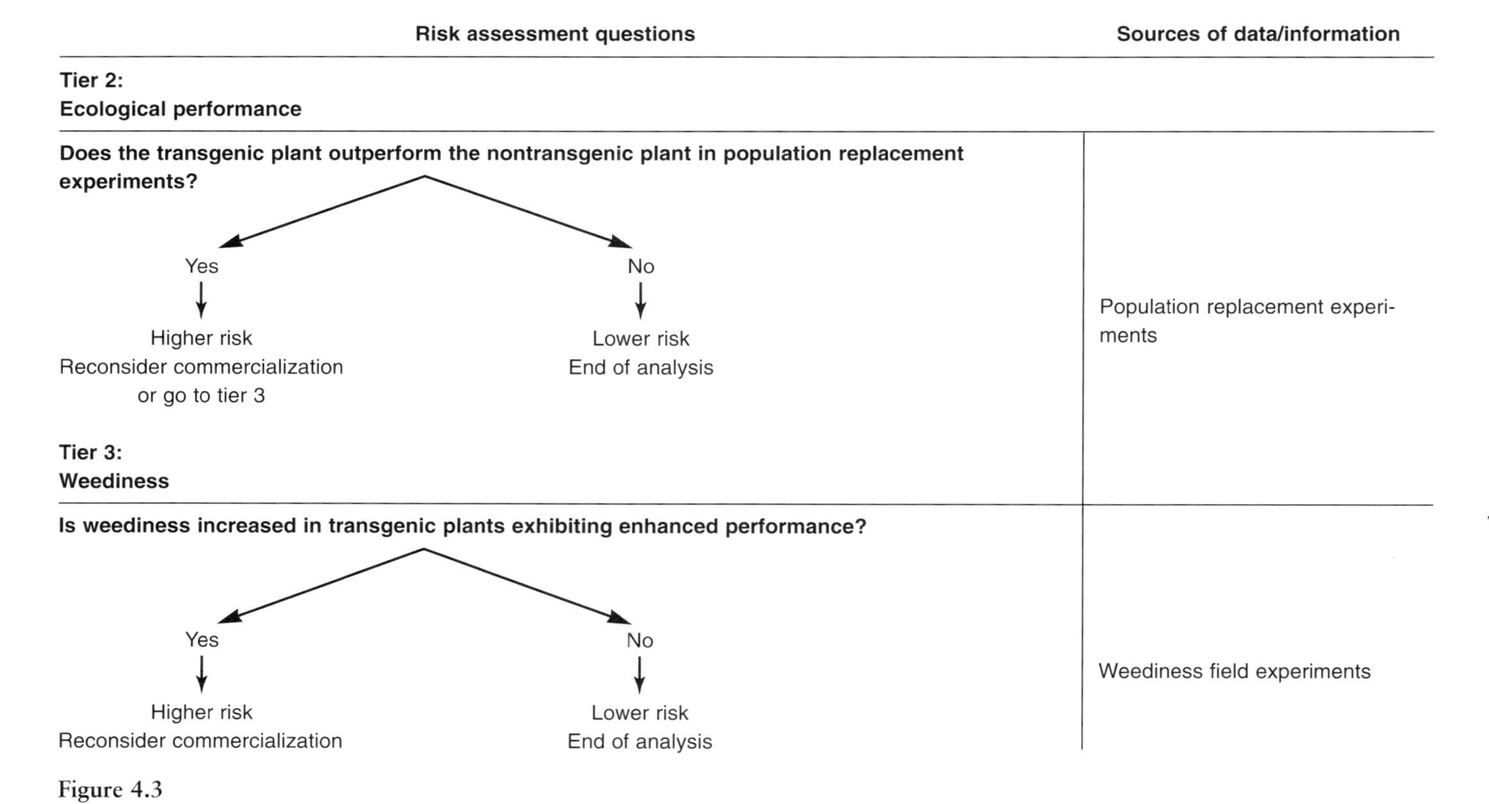

Figure 4.3
Summary of scheme to assess two environmental risks

The fundamental question addressed by the testing scheme is whether the addition of transgenes to crop plants by genetic engineering techniques or to wild/weedy plants by gene flow changes those plants into weeds. The testing scheme focuses on only one element of weediness—the capacity for a transgenic population to invade ecosystems. Two three-tiered analyses were developed to answer the question. In both analyses, the tiers are designed to identify less risky plants early and to require extensive field testing only for those plants that appear to pose substantial risks.

The first-tier assessment in both cases is based, for the most part, on existing data. In terms of crop weediness, tier 1 uses information on the nonengineered crop and its relatives to place a transgenic crop into one of two categories: low or high weediness potential. This categorization determines the level of testing that a crop will undergo in tier 2. Crops with higher weediness potential are subject to a standard set of tier 2 experiments, whereas crops with lower weediness potential undergo an abbreviated set of experiments.

In evaluating gene flow, the first tier determines whether a crop can form viable, fertile hybrids with any relatives that exist in the United States. In other words, tier 1 reveals whether genes can flow from the transgenic crop to wild/weedy relatives. Only those that form viable, fertile hybrids are subject to further testing in tier 2.

Tier 2, which is generally the same for transgenic crops or transgenic wild/weedy relatives, requires that experimental data be generated on the plants themselves. The data, derived from relatively simple population replacement experiments, indicate the potential for transgenic plants to become invasive weeds. Population replacement refers to the capacity over time of a population of plants to produce viable, fertile offspring. It is a measure of the increase or decline of a genetic type over generations—that is, whether a genetic type will persist over time. This measure encompasses two pieces of information: the rate at which a population replaces itself and the persistence of its seed bank.

Population replacement data are used to compare the "ecological performance" of transgenic versus nontransgenic plants. For example, if the experiments show that the population of engineered plants declines and its seed bank persists less well relative to the nontransformed counterpart, then it is highly unlikely that the transgenic plant will be able to cause greater adverse impacts than the nontransgenic plant. Generally, an organism that is maintaining itself less well in the environment than a counterpart would not be expected to pose greater risk than that counterpart over time.

After the second-tier analysis, crops either are eliminated from further analysis or are sent into the third tier. The third tier offers an opportunity for developers to show that transgenic plants that outperform nontransgenics in tier 2 tests do not pose risks as weeds.

Most transgenic crops developed for the United States will likely fall into lower-risk categories by the end of the tier 2 analyses. Thus, most crops will need to undergo only three years of testing. In some cases, tier 2 testing may require four or five years, depending on how many generations of wild/weedy relatives are required to evaluate gene flow. It is expected that the tier 2 experiments can be conducted along with efficacy tests that companies already perform.

Where crops fall into the higher-risk category and companies choose to pursue tier 3 testing, additional years of experimentation can be anticipated.

Other risks identified in chapter 3 remain difficult to evaluate. First, the lack of information on the distribution of wild relatives in many parts of the world makes it difficult to evaluate gene flow outside the United States. Second, scientists have barely begun to consider the long-term, cumulative risks to ecosystems from transgenic crops. Predictions based on anything other than broad, general principles will be unavailable for quite some time.

In the near future, scientists may be able to devise ways of assessing two risks identified in chapter 3. It may be possible to develop a scheme for identifying certain virus-resistant crops that pose less risk than others. Virologists may also be able to design methods that either reduce the risks of certain crops or help to identify problems in others. Using methodology developed for other purposes, scientists may be able to devise protocols for assessing the nontarget impacts of pesticides and pharmaceuticals produced in transgenic crops. Nevertheless, there is a clear need for aggressive research programs to narrow the uncertainties and support the goal of science-based risk assessment.

5

International Implications of Commercialization

Seed, pesticide, and biotechnology companies are already positioned to introduce the new transgene technology globally. Transnational companies—Cargill, Monsanto, Hoechst-Roussel, and others—are racing to develop engineered crops for a worldwide market.

Many companies own subsidiaries or operate branches and research stations around the world. For example, Pioneer Hi-Bred, one of the world's seed giants and a major U.S. investor in crop genetic engineering, has wholly or partially owned subsidiaries in thirty-one countries—in Africa, Asia, Australia, Europe, the Middle East, and North and South America (*Seed World*, 1992). Over the next few decades, farmers around the world will be urged to buy seeds of crops engineered by these companies.

Other agricultural biotechnology companies have established joint research and marketing ventures with foreign companies or with U.S. companies with foreign interests. Agrigenetics, for example, had research and marketing arrangements with international companies based in Argentina, France, and Italy (*Seed World*, 1992). Calgene signed agreements with U.S. growers and packers to produce its genetically engineered tomatoes in Mexico as well as the United States (*Biotech Patent News*, 1992). Monsanto has donated its transgenic potato technology to Mexican scientists (Gershon, 1992).

A Global Seed Trade Means Global Risks

Engineered crops, whether they are grown in the United States or abroad, are potentially harmful to the environment. The risks they present worldwide are generally the same as those identified for the United

States in previous chapters: gene flow to crop relatives; weediness in all its aspects—cumulative and cascading effects, nontarget impacts, secondary effects on agricultural ecosystems; and creation of new viral strains or viruses with broadened host range.

Many of these consequences, such as new weeds, will be local in impact and thus primarily the concern of the country involved. But other effects have global significance. Of particular concern from the U.S. perspective are threats to centers of crop diversity: areas that support populations of traditional crop varieties (landraces) and their wild relatives. These plants are the genetic basis of the world's future food supply. They are the source of new genes that plant breeders and genetic engineers use to adapt crops to changing environmental conditions. As the accompanying map (figure 5.1) shows, most of the centers of diversity are in developing countries.

Diversity in the centers is already disappearing at an alarming rate, however, because farmers have abandoned landraces in favor of Green Revolution cultivars, and habitats are being destroyed as the human populations expands (table 5.1). Fowler and Mooney (1990) document, for example, that indigenous stands of wheat have virtually disappeared from India and Greece since those countries began intensively planting Green Revolution varieties. In North Africa, native wheats of the Nile Valley are being replaced by modern varieties. And in Nepal, new varieties are currently grown on some 80 percent of the land—displacing roughly 80 percent of native varieties. As Fowler and Mooney (1990)

Table 5.1
Loss of crop varieties—four examples

Crop/Area	Lost	Remaining
Rice varieties/India	30,000	12[a]
Rice varieties/Southeast Asia	up to 100,000	1[b]
Vegetable and fruit varieties/United States	97%	3%
Wheat varieties/Middle East	85%	15%

a. Covering 75 percent of rice fields
b. Covering 65 percent of paddy land
Source: Adapted from "The Seed Map: Dinner on the Third World," Rural Advancement Foundation International, Pittsboro, N.C., 1992.

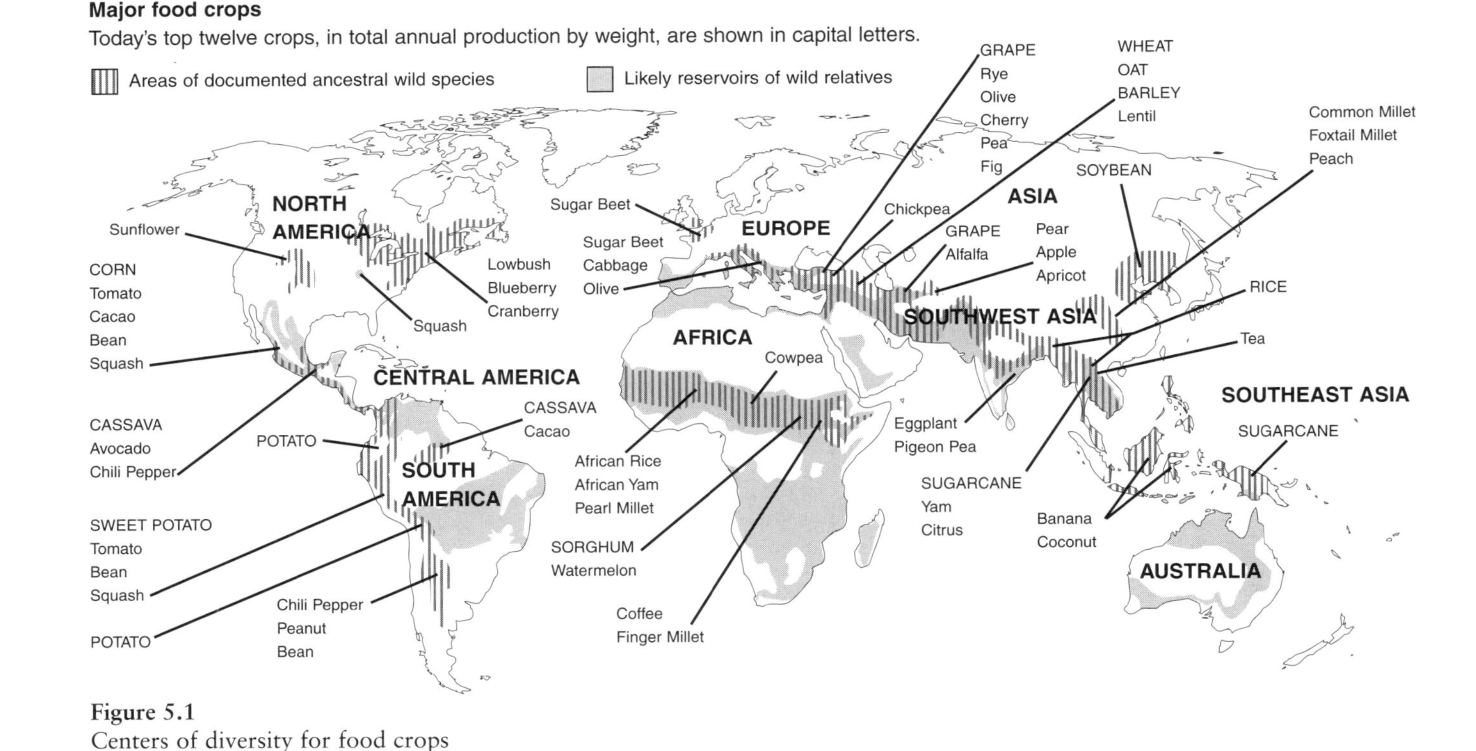

Figure 5.1
Centers of diversity for food crops
Sources: Jack R. Harlan, Professor Emeritus, Plant Genetics, University of Illinois; and a map produced by the National Geographic Society, *National Geographic*, April 1991.

Centers of Diversity

Centers of diversity are regions around the world harboring populations of relatives of crops. These populations constitute a reservoir of genetic material that can be moved into the crops by traditional breeding techniques. For example, relatives of corn are found in Mexico, which is the center of diversity for corn. Diversity in the centers may be found in the form of wild relatives of the crop or traditional crop varieties, also called landraces (Fowler and Mooney, 1990; Rural Advancement Foundation International, 1992). Some landraces have been used by native peoples for centuries.

Most crops have only one or a small number of centers of diversity. For example, corn, soybeans, and potatoes currently grown around the world have few relatives in most areas where they are cultivated. Rather, their relatives are found in a few areas where the crop has been grown for millennia.

Areas of greatest diversity often, but not always, coincide with the centers of origin, that is, where crops were first domesticated by humans (Fowler and Mooney, 1990; Rural Advancement Foundation International, 1992). For example, Mexico is both the center of origin and center of diversity for corn. Ethiopia, on the other hand, is a center of wheat diversity but not the center of wheat origin, which is the Middle East.

Centers of diversity are essential to the survival of world agriculture because the landraces and wild relatives provide the raw genetic material for breeding new characteristics into crops. Breeders in need of new traits go to centers of diversity, where they may look for qualities like disease resistance, cold tolerance, and drought resistance (Fowler and Mooney, 1990; Juma, 1989). For example, to add disease resistance to a crop, breeders turn to wild relatives or landraces that exhibit resistance, and attempt to breed those traits into crops. Crop stocks may eventually decline in value if they are confronted with new environmental stresses to which they cannot be made resistant for lack of new genes from relatives.

show, similar effects can be seen around the world in crops including rice, potatoes, millet, sorghum, and many others.

Widespread use of engineered crops may exacerbate this loss of diversity in two ways. First, as described in chapter 3, wild relatives may be displaced by crops or other populations carrying advantageous transgenes. This is particularly important in communities where small populations of relatives may be the lone repositories of certain genes, as in the case of some relatives of corn that occur in only a few sites in Mexico (Iltis and Doebley, 1980). Moreover, transgene flow to landraces, with subsequent human selection for those carrying advantageous traits, could diminish diversity in the traditional varieties as well.

> The ethical imperative should therefore be, first of all, prudence. We should judge every scrap of biodiversity as priceless while we learn to use it and come to understand what it means to humanity.
> —E. Wilson, 1992, p. 351

Second, the pressure to replace landraces with new cultivars may be intensified as the agricultural biotechnology industry markets transgenic cultivars of major and minor crops around the world. For the most part, landraces abandoned by traditional farmers face extinction (Brown and Young, 1990; Fowler and Mooney, 1990; Rural Advancement Foundation International, 1992; Young, 1990). Taken together, these two risks may well accelerate the already dramatic loss of the genetic basis of the world's food supply.

U.S. Approval of a Transgenic Crop Does Not Assure Global Safety

Although risks of transgenic crops are generally of the same kind no matter where they are grown, the level of risk associated with a particular crop will vary from country to country, depending on the nature of the crop itself and the environment in which it is planted. For example, engineered insect-resistant soybean may pose minimal risk in the United States because there are no sexually compatible wild relatives in this country. They would likely be approved for use in the United States where the review would assume use in this country. That approval, however, would say nothing about growth of the soybean outside the United States. In fact, cultivation in China, where large numbers of landraces and wild relatives of soybean virtually ensure that transgenes will move to nearby relatives, may pose serious risks.

> If the diversity of both crop and noncrop species is not safeguarded, much of the raw material now available for genetic manipulation will be lost. Biotechnology can move genes, but its ability to create them is virtually nonexistent.
> —L. Brown and J. Young, 1990, p. 72

As the example illustrates, an assessment of risk in the United States will not take into account all the environmental variables confronted by a crop used in different countries in different parts of the world. Other countries cannot rely on the United States for assessment of global safety. Other entities must perform assessments of the risks entailed in the use of transgenic crops in other countries. Ideally, each country would assess the transgenics and weigh the risks against the benefits.

Why Worry about Reducing Biological and Genetic Diversity?

In the last few years, the loss of biodiversity has become a major subject of international conferences and in the popular and scientific press (*AMBIO*, 1992; Fowler and Mooney, 1990; Mann and Plummer, 1992; Wilson, 1992). Typically, discourse on biodiversity focuses on the alarming rate of extinction of entire species, the major role of habitat destruction in accelerating that rate, and reasons why humans should reverse those trends to save endangered spotted owls, butterflies, wild orchids, and tropical trees.

Beyond the loss of entire species, the disappearance and impoverishment of local ecosystems threaten biodiversity. In fact, Ryan (1992) and Ehrlich and Dailey (1993) suggest that populations of genetic variants within species may be vanishing at a higher rate than whole species.

People worry about this loss of biodiversity because of its threat to the ecological services on which humanity's existence depends. These services may be direct ones, that is, providing food, fiber, building products, and medicines. Or they may be indirect—the services that make up the earth's life-support system. These include converting sunlight to plant carbohydrates; recycling nutrients and water; maintaining the mixture of gases in the atmosphere; generating and maintaining soils; and controlling pests (Ehrlich and Ehrlich, 1992).

Some people also value organisms aside from their importance to humanity's survival. These values may be esthetic, focusing on the pleasures associated with natural beings, or ethical, dealing with the intrinsic right of other life forms to exist (Ehrlich and Ehrlich, 1992).

Extinctions also mean that the genetic diversity needed to adapt to changing environments is diminished. In the centers of crop diversity discussed above, this means the loss of genes that breeders might use to protect and fortify the world's food supplies. Moreover, these losses shatter intricate links between populations, communities, and ecosystems. These localized insults, repeated countless times around the globe, eventually weaken the earth's unique life-sustaining system.

Unfortunately, it appears that most countries are not prepared to control the ecological risks of transgenic crops. Countries where most centers of diversity are found are among those least likely to have the resources needed to protect against the risks of the technology (United Nations Industrial Development Organization, 1992). Because of the importance of centers of diversity to world agriculture, they may deserve protection under international biosafety protocols that may be developed for genetic engineering.

In the meantime, it is vital that approval in the United States not be mistaken as a global seal of safety. Ecological risk assessments, unlike chemical toxic assessments, for example, do not travel. It is incumbent upon the United States, as a major promoter of biotechnology, to see that this misconception does not take hold.

Teosinte and Corn: A Cautionary Tale

As an example of the potential international impact of engineered crops on wild relatives, let us consider a new cultivar of corn genetically engineered to produce a foreign gene for an insect toxin that would repel pests. Under the scheme described earlier in this book, the U.S. government would examine the corn for two major environmental risks—weediness of the corn plant itself and the impact of the transfer of the foreign genes into wild plant relatives.

Corn is unlikely to present serious risks in either category. Domestication has robbed corn of many of the traits necessary for survival outside of the confines of cultivation. Corn seeds, for example, are not readily dispersed away from the parent plant. The field tests conducted under the scheme described above would likely confirm the expectation that newly pest-resistant corn still could not survive on its own and would be unlikely to become a weed.[1]

As long as the analysis is restricted to the United States, gene transfer is also unlikely to be a problem for U.S. regulators. Few relatives of corn grow within U.S. borders.[2] Without relatives growing in the vicinity of the crop, no gene transfer can occur. Presuming it posed no other hazards—such as nontarget effects, food safety, or other concerns—corn engineered to be pest resistant would likely be approved for use in the United States.

Although the analysis is limited to the United States, the use of the seed need not be. In fact, once approved in this country, the seeds would be offered for sale in the international seed market, and could easily be purchased by farmers in Mexico or Guatemala. These two countries harbor teosinte, the wild grass from which corn is believed to have originated.[3] Restricted to a range in the southern and central parts of Mexico and Guatemala, teosinte often grows in and around corn fields (Wilkes, 1977). In fact, its distribution parallels some of the finest agricultural land in Mexico (Wilkes, 1977). Corn and teosinte produce fully fertile hybrids (Wilkes, 1977) and are known to exchange genetic material freely and continuously under field conditions (Doebley, 1990; Wilkes, 1977).

Thus, pollen containing the foreign pesticidal toxin gene would readily move into wild teosinte populations, where it could have adverse effects. If the toxin were to make the weeds more resistant to pests, for example, it could give one subpopulation of teosinte an advantage over another. In a worst case scenario, the gene could lead to the extinction of some teosinte subpopulations and accelerate the reduction in diversity in these important plants. Teosinte populations are already declining at rates such that they could become endangered (National Research Council, 1993). Of the eight distinct population clusters of annual teosinte, three are considered rare, occurring at single locations (National Research Council, 1993).

In some circumstances, even if the pesticide-containing corn were not advantageous to wild populations, its genes could still be retained in the populations and do harm. Where the populations of teosinte are small compared to the engineered corn crops, the pollen from the engineered corn could simply swamp the wild plants with so much pollen that all progeny would be hybrids containing genes from the engineered plants. If the engineered pollen contained traits deleterious to wild teosinte, it could lead to its extinction.[4]

So, who would care if small populations of teosinte were extinguished on some remote hillside in Guatemala? Certainly the Mexicans and Guatemalans would.[5] But the United States also relies on wild relatives of corn to replenish and maintain the vigor of one of its most important crops. Currently, the practice of planting millions of acres in genetically uniform varieties of corn has made U.S. agriculture extraordinarily vulnerable to pests. If a pest is able to break through corn defenses, practically the entire crop is at risk. In that case, scientists will turn to other corn cultivars and wild teosinte populations for resistance traits. If too many of the small populations of teosinte become extinct, agricultural breeders may not be able to find the resistance trait needed to save the corn crop and the United States could pay a terrible price.

Such a scenario is not as farfetched as it might sound. In 1970, the U.S. corn crop in the South and the Corn Belt was devastated in a few short months by a single fungus (Doyle, 1985). The fungus, which caused a disease called the southern corn leaf blight, was able to move swiftly from one corn field to another because almost all the corn was susceptible. The quest for high yields had led farmers to rely on a few genetically uniform cultivars of corn which were susceptible to the disease. No pesticide saved the day. Resistance had to be found either in other corn cultivars or wild corn relatives. Breeders found the resistance genes in the 1970s. Will they be able to do so the next time?

How can teosinte be protected from the risks posed by engineered corn? Doebley (1990) concludes that the risk can be controlled by not growing engineered corn in the region where teosinte is native. Thus, the most effective and straightforward approach would be to bar engineered corn from the restricted geographic regions where teosinte is found. Alternatively, Mexico and Guatemala could subject new engineered corn strains to testing regimes similar to ones described earlier in this book, adapted to their own situations.

The questions U.S. regulators must address are whether Guatemala and Mexico have programs in place to protect the teosinte populations and whether those countries will have sufficient notice of the introduction of engineered corn to initiate regulatory actions. Because of the leading U.S. role in the development of transgenic crops, and because of the global significance of centers of diversity, the United States should initiate international efforts to assist in their protection. In particular, the United States should begin to determine what kinds of assistance it could offer to help countries with diversity centers to undertake protective measures.

6

Conclusions and Recommendations

In this book we find that the commercial-scale uses of some engineered crops could pose serious ecological risks. Transgenic crops may become weeds in farmers' fields or they may invade neighboring areas, ultimately altering local ecosystems. It is also possible, but generally less likely, that widespread use of transgenic virus-resistant crops could result in new virus strains or viruses with expanded host ranges.

Through the flow of pollen, some transgenes may end up in populations of wild or weedy relatives of the crop. Some of these transgenic relatives may become new weeds that farmers must control. Other wild/weedy relatives that are the recipients of transgenes may alter the makeup of the communities and ecosystems of which they are a part.

When transgenic crops are planted in regions harboring relatives and traditional varieties of crops (centers of diversity), the significance of gene flow increases. In these centers, the movement of transgenes could threaten the diversity of landraces and wild/weedy populations. These are plants vital to the future world food supply, as they are the source of genes that allow plant breeders to modify crops as environmental conditions change.

We have also found that, while some ecological risks are difficult to evaluate, others can be assessed. We have suggested a novel approach to evaluating weediness—one that separates lower-risk from higher-risk transgenic crops on the basis of empirical data collected on the new crops.

Other aspects of weediness are difficult for government regulators to evaluate in a meaningful way. First, because of the lack of knowledge of local flora in many parts of the world, the risks of gene flow to wild relatives cannot be assessed in many countries. Second, long-term and cumulative impacts depend on the dynamics of ecosystem interactions through time that at present are not well understood. Third, pharmaceutical- and pesticide-producing plants pose special risks because of the toxicity of the transgene product. Using methods developed for other purposes, scientists may be able in the near future to develop protocols for analyzing the nontarget effects of these transgenic plants. Finally, risks of virus-resistant plants, some of which are fairly well understood, cannot be predicted because key components of the methodology are lacking. All these issues indicate the need for a strong regulatory regime.

The Need for Regulation

Few people are comfortable with the notion that scientists should be able to construct engineered organisms of any kind and introduce them anywhere in any quantity. There is broad agreement among governments around the world, including the United States, that the technology should be regulated for its environmental risks. The major issues are how extensive the regulatory oversight should be, whether new legislation should supplement or replace existing laws, and how effective regulation of international sales may be achieved by international agreements.

Because of the diversity of products and activities expected from the technology and the large number of potentially applicable statutes, the regulation issue is a complicated one. At present, the United States has no comprehensive statute addressing the environmental risks of transgenic organisms. Instead, products and activities of biotechnology are covered under a patchwork of existing statutes. In the existing framework, transgenic plants are potentially regulated under statutes governing plant pests, food and drugs, and pesticides—yielding a confusing mix of up-and-running programs, pending regulations, and unresolved policy questions. Uncertainties that were tolerable while activities were confined to field testing are becoming acute as products are being commercialized.

Since 1987, the USDA has overseen the field-testing and commercialization of genetically engineered crops under a program implementing the Federal Plant Pest Act. The program requires either prerelease notice or approval of all field trials of genetically engineered crops (U.S.

Department of Agriculture, 1987, 1993). Once approval to test has been obtained, there is no requirement that companies use the field-test period to conduct experiments of the kind discussed in this book. In fact, most sponsors do not collect ecological data on their crops during the field-testing period (Mellon and Rissler, 1995).

Before commercialization, the USDA prepares and makes available for public comment comprehensive analyses of the environmental risks posed by the unrestricted commercial use of the crop. Most of the information for the analyses is provided by sponsors and the USDA. Lacking experimental data, the USDA relies primarily on armchair analyses—extrapolations from the scientific literature to assess the risks of transgenic crops.

In this book we offer a risk assessment scheme that can be adopted by the USDA under its current program. Under this scheme, the experimental evaluation of ecological risks would be accomplished during the company's four to six years of field testing.

It should be emphasized that we address only certain types of environmental risks associated with transgenic crops; a product that is intended for use as food, for example, may require additional review before it ends up on the grocery shelves.

The Role of the Public

As in so many technological issues, the public has a vital role to play in the development and direction of biotechnology. Public education about the technology is needed to enable people to evaluate its risks and to participate in decisions about its use. The public education needed is not simply a campaign to calm fears and render the technology acceptable to the public. Doubts about the technology do not come only from unfounded fears born of too many late-night horror shows. Unreasonable fears exist and should be countered, but there are numerous genuine risk and nonrisk concerns connected with the advent of the technology.

An educated public can evaluate the risks and weigh them against the benefits. Just as important, the public should have a say in choosing among the benefits of the technology. In agriculture, for example, public participants should help decide between products and research that support sustainable agriculture and those that support chemical-intensive systems.

Biotechnology must be evaluated application by application. Generally speaking, with proper controls the technology should prove useful in basic scientific research and in commercial fermentation systems to produce drugs and specialty chemicals. For example, use of engineered bacteria to produce substances, like insulin, that are otherwise unavailable in needed quantities is to be welcomed. As for the environmental applications of the technology, they should be approached cautiously until more is known about both risks and benefits. Environmental releases of engineered organisms, particularly on a commercial scale, pose potentially significant risks. Thus, caution is needed as well as strict government oversight of planned introductions.

In agriculture, some of the most promising and exciting developments for the future are in sustainable agriculture. Biotechnology products could play a positive role in the transition to an environmentally sound agriculture, but guarantees are needed to ensure public involvement in selecting among particular applications of the technology. Not all applications will prove beneficial, and society need not regard technology as a sort of train that cannot be stopped once it gets on track. We can and must learn to evaluate technologies and, where appropriate, to say no.

Finally, a well-developed public research effort is needed to complement private research and development efforts. Publicly funded programs, freed from commercial constraints, can investigate innovative approaches to problems, such as crop rotation to control pests, that depend on skilled management rather than product inputs. Publicly funded research can also develop products, such as narrow-spectrum biopesticides, that do not offer a sufficient return to justify private investment for research and development. Such programs also provide opportunities for public involvement in setting the direction of technological advancement.

It would be unwise to brush aside concerns about genetic engineering in headlong pursuit of its benefits. Cautious development of the technology will give us the opportunity to avoid its risks and insure that it serves—not dictates—our social and environmental goals.

Recommendations

To protect against the ecological risks of transgenic plants, the Union of Concerned Scientists makes the following recommendations:

1. The United States should strengthen the federal regulatory program to include a strong experimental basis for the assessment of the risks of transgenic crops.

The program should consider the ecological risks to agriculture and natural ecosystems in the United States and elsewhere in the world and should pay particular attention to the protection of the world's centers of diversity for important food and fiber crops.

The program should be administered by an agency with an environmental mission. The environmental risks of the release of transgenic plants should be regulated by an agency with a mandate to protect the environment. The agency selected should have strong, clear authority to control commercial activity to protect the environment and act in emergency cases. Currently, the USDA, the primary overseer of transgenic crops, lacks the needed mandate.

The program should evaluate each transgenic crop before it is approved for commercial sale. Each crop/transgene combination is different and may present different risks. Until a substantial body of data is collected on the ecological impacts of a wide range of transgenic crops, each combination warrants a separate assessment, although its extent may vary from one transgenic crop to another. Categorization of some crop/transgene combinations into a low-risk category may be possible in the future. In view of the uncertainty about long-term risks, approval decisions should err on the side of caution. The long-term impacts of transgenic plants in the environment are unpredictable and difficult, if not impossible, to predict at the current level of scientific understanding. Yet these impacts could cause systemic and generally irreversible changes in ecosystems. Regulators should therefore be cautious in allowing commercialization of transgenic crops having unproven environmental benignity.

2. All transgenic crops should be evaluated for at least two aspects of ecological risk—weediness potential and gene flow—before they are approved for commercialization.

Effects on centers of diversity within the United States should receive special attention. Some transgenic crops pose substantial ecological risks, which, if not controlled, could cause significant losses in agriculture and genetic diversity. As this book shows, two major features of ecological risk can be reasonably evaluated and therefore should be assessed before commercialization. In addition, human and animal health risks must be assessed before the crops enter the marketplace.

The federal regulatory program should employ a standard risk-assessment scheme, data requirements, and public review and comment processes. Unlike current regulation, decisions should be based on experimental data gathered during the field testing of the crop rather than on armchair analysis. The review process should incorporate public involvement before any decisions are made to approve or deny applications for commercialization.

This book outlines a general approach to the risk assessment of weediness potential and gene transfer. The suggested approach separates higher-risk transgenic crops from lower-risk crops using empirical data on the field behavior of transgenic in comparison to nontransgenic crops.

3. The federal government should develop standard protocols to assess the risks of creating new viruses, nontarget effects of pesticides, and the ecotoxicity of plant pharmaceuticals.

Virus-resistant crops are likely to be a major product of agricultural biotechnology. The possibility that widespread use of such plants might lead to the creation of new viruses, or other harmful effects, is now being widely discussed. We recommend that experiments be done to answer the question of whether rates of recombination in transgenic plants are greater or less than those in nontransgenic plants infected with two or more viruses. If experiments indicate that they are significantly greater, protocols should be developed to identify, assess, and screen out harmful new strains.

Both pesticide- and pharmaceutical-producing plants involve the introduction of potentially toxic substances into the environment. Although the problems these plants might cause are theoretically covered under the broad definition of "weediness," the use of potentially toxic substances in crop plants raises enough special issues that tailored protocols are needed.

4. The U.S. government should sponsor research that would enable a full assessment of all ecological risks of genetically engineered crops.

Many questions related to the risks of transgenic crops need systematic research before the risks can be adequately predicted and controlled. Currently, the federal government administers a small program of research grants in the area of environmental risk assessment. This program, however, is woefully inadequate to produce the systems-level data and methodology needed to ensure the safe release of engineered organisms.

The United States should sponsor more ecosystem and risk assessment research to fill this gap. The research programs of the USDA, the Environmental Protection Agency, the Department of Commerce, and the National Institutes of Health should include risk assessment and related research as priority objectives.

5. Congress should direct the National Academy of Sciences to prepare a report on (i) the likelihood that seeds of engineered crops developed in the United States will be dispersed to centers of crop diversity, and (ii) the availability of floristic surveys and other information needed to assess the impacts of engineered crops released in countries harboring the centers.

The United States has an interest in protecting landraces and wild relatives of crops wherever they exist. These plants are sources of genetic diversity on which the future viability of crops, including those grown in the United States, depend. Biotechnology cannot create new genes; it can only find and move them. Thus, the genetic diversity is important for genetic engineers and traditional crop breeders alike.

Centers of diversity are already eroding under pressure from loss of habitats and the tendency of modern agriculture to rely on a few elite varieties of important crops. Hundreds of thousands of varieties of crop relatives have been lost. The U.S. government, however, shows no inclination to assess risks posed in other parts of the world by crops engineered in the United States. At the same time, no international or country-by-country system exists to assess risks of commercial releases of engineered crops, nor do developing countries—which possess most of the centers of crop diversity—typically have the resources to assess risks of engineered crops that may reach their farmers.

The U.S. government should help develop a protective policy by evaluating the movement of seeds as a result of the global seed trade and the floristic data and other relevant information needed to examine the risks of engineered crops in centers of diversity.

6. All transgenic seeds that are exported from the United States should bear a label stating that approval of the seeds under U.S. law carries no implication of safe use in other countries.

The risk assessment approach described in this book is dependent on specific information about crops, crop relatives, and weed populations grown or found in a particular location. Thus, a finding that a transgenic crop can be grown safely in the United States cannot be automatically applied to other countries.

For example, it might be safe to grow transgenic corn in the United States because there are few wild relatives of corn here. The same corn planted in Mexico might not be safe if it were planted in the region where dwindling populations of wild corn relatives are found. These small populations, which are the major reservoir for genetic variety for corn, should be protected from the flow of new genes coming from transgenic corn.

It is, of course, up to the Mexican government to perform the analyses and decide how to respond to a threat where one exists. The United States can aid in the protection of important populations like these by making it clear through labels that the U.S. safety assessment may not be applicable to Mexico and cannot be relied on by the Mexican government as an assurance of safety.

7. The United Nations should develop international biosafety protocols, which are necessary to ensure that developing countries, especially those harboring centers of crop genetic diversity, can protect against the risks of genetically engineered crops.

Most countries around the world are unprepared to evaluate and control the risks of transgenic crops. Developing countries, where most centers of diversity are found, are among those least likely to have the resources to protect against the risks of the technology. Multilateral efforts are needed to develop biosafety guidelines that will enable these countries to become fully informed about and to evaluate and minimize the risks of genetically engineered crops.

Appendix: Experimental Assessment of the Replacement Capacity of a Population of Plants

There are a number of ways to evaluate the replacement capacity of a genetic type in a population of plants. Option A describes one alternative. N. Ellstrand, J. Hancock, P. Kareiva, R. Linder, R. Manasse, and M. L. Roush were especially helpful in developing this experimental approach. Another option, offered in B, was developed by R. Manasse. (See Linder and Schmitt, 1994, for another approach.)

Option A

Purposes

- Determine net replacement rate (R) values for both transgenic (t) and nontransgenic (n) plants in a three-generation experiment.
- Determine half-life (H) of transgenic and nontransgenic seed bank.

Year 1

Net Replacement Rate Experiment

Sow a known number of transgenic and nontransgenic seeds randomly in plots (for example, 1 m^2) in blocks.[1,2,3,4]

At end of year 1:

i. Collect seed samples from each plot.

ii. Determine number of viable[5] n and t seeds[6] from each plot.

iii. Calculate R for each t and n plot in each environment; for example, for a t plot, R = # viable t seeds collected ÷ # viable t seeds sown.

iv. Calculate R_t and R_n for year 1 as average of R_t's and R_n's from t and n plots.

v. At end of data collection, sow remaining seeds in plots from which they came.

Seed Bank Experiment

Place a known number of transgenic and nontransgenic seeds in rot-resistant bags, and bury at systematically varying depths in blocks as above.

- Determine viability at time of burying.
- At six months and end of year, remove sample bags and determine number of viable seeds remaining in *t* and *n* seed banks for each environment.

Year 2

Net Replacement Rate Experiment

At end of year 2:

i. Collect data as at end of year 1.
ii. Calculate R_t and R_n for year 2.
iii. Sow seeds as at end of year 1.

Seed Bank Experiment

At six months and end of year, remove sample bags and determine number of viable seeds remaining in *t* and *n* seed banks.

Year 3

Net Replacement Rate Experiment

At end of year 3:

i. Collect data as at end of year 1.
ii. Calculate R_t and R_n for year 3.
iii. Calculate overall R_t and R_n over the three-year period for each environment.

Seed Bank Experiment

At six months and end of year, remove sample bags and determine number of viable seeds remaining in *n* and *t* seed banks.

Calculate H_t and H_n. Assuming a negative exponential decay rate, the half-life is estimated by taking the logarithm of the actual number of seeds remaining and performing a linear regression on the values with respect to time. The regression produces a line from which can be estimated the half-life.

Option B

Rather than determine the replacement rate as in Option A, one can measure the rate of change in population size relative to the competitor. In year 1, the experimenter plants equal numbers of transgenic and nontransgenic seeds at a rate that will produce a number of mature plants near the carrying capacity of the plot.

At the end of the growing season, the seeds are collected and the genetic types determined. The rate of replacement of the transgenic relative to the nontransgenic is the frequency of the transgenic type produced divided by the frequency of that type planted. For example, given that the starting frequency of the transgenic was 0.5, if after one year the frequency in the collected seeds is 0.6, the relative rate of replacement is 1.2. The relative rate of replacement of the nontransgenic is $0.4 \div 0.5 = 0.8$.

In year 2, and in the following year, the beginning number of seeds remains the same but the relative proportions of the genotype differ and are equal to the proportions from the end of the previous year. If the proportion of seeds from the previous year with the transgene gene was 0.6, then 0.6 of the seeds sown in the next year should have the transgene.

The important number is the rate of decline (or increase) after three years (where rate is calculated per three years). The relative rate of change after three years is the product of the rate calculated each year, or simply the outcome (the rate) after the third year.

At the beginning of the experiment, the competition is equal and the relative rate of replacement is 1.0. If, at the end of the experiment, the relative rate of replacement for the transgenic variety is greater than 1.0, then it can be inferred that that genotype is outperforming the nontransgenic.

Glossary

Abiotic stress Nonliving factors that may cause a deleterious effect on plant. Examples include pesticides, temperature and moisture extremes, and soil and water salinity or acidity.

Agroecosystem The organisms and abiotic factors with which they interact in a field or other portion of an agricultural enterprise.

Allele Alternative versions of a gene. Genes may exist in two or more forms. For example, a gene for height in peas may exist in two versions: one allele for short peas, the other allele for tall peas.

Asexual reproduction Means of reproduction that does not involve sex. Offspring of asexual reproduction are identical to the parent. Asexual propagules, or vegetative parts capable of reproducing the plant, are rhizomes (potato), runners (strawberry), stem pieces (willow), and seed (dandelion); also called vegetative reproduction.

Backcross Crossing a hybrid with a member of one of the parent population.

Biodiversity The vast array of the earth's organisms and their genes. Embedded in the concept is the interrelatedness and interdependence of genes, organisms, communities, and ecosystems.

Biotechnology Use of living organisms for human purposes. Genetic engineering methods—gene splicing or recombinant DNA methods—are one set of techniques used in modern biotechnology.

Biotic stress Living organisms that may cause a deleterious effect on plants. Examples include viruses, bacteria, and fungi that cause disease and insects that feed on plants.

Centers of origin and diversity Places in the world where crops have the greatest genetic diversity in the form of traditional crop varieties and/or wild relatives. Centers of diversity are typically, but not always, the same locations as the centers of origin or oldest cultivation of the crop.

Chromosome Structures along which genes are located.

Coat protein Outer envelope of protein that encloses the nucleic acid core of a virus; also called capsid.

Coat protein-mediated protection Protection of a plant against infection by a virus, obtained by splicing into the plant genome a viral gene for coat protein from a similar virus.

Community An assemblage of populations of organisms that interact.

Cross-pollination Pollen from one plant transferred to the female part of a plant of a different genetic makeup.

Defective virus Form of a virus lacking the capacity to fulfill one or more critical biological functions.

Dormancy Period in which growth does not occur; growth begins only when certain environmental requirements, such as temperature and moisture, are met.

Ecological performance A term adopted in this book to refer generally and in nontechnical language to the overall results of population replacement experiments for a particular crop.

Ecosystem The organisms in a community and the abiotic factors with which they interact.

Encapsidation Process by which viral nucleic acid is enclosed in coat protein (capsid).

Epistasis Process in which one gene modifies the expression of another gene that is not an allele of the first.

F_1 generation Generation of hybrids from the cross of two parents differing from each other in one or more genes. F_1 refers to first filial generation.

F_2 generation Generation of plants produced after interbreeding of hybrids or F_1 generation; F_2 refers to second filial generation.

Fertilization Union of egg and sperm (pollen) to produce offspring.

Floristic survey Listing of the species of plant life (primarily flowering plants) in a particular area.

Gene Functional unit of heredity usually carried on the chromosome and made up of DNA.

Gene flow Exchange of genes between different, usually related, populations. Genes commonly flow back and forth among plants via transfers of pollen.

Gene pool Sum of all versions (alleles) of all genes of all individuals in a population.

Gene splicing Combining genes from different organisms into the chromosomes of one organism; also called recombinant DNA techniques.

Genetic assimilation Dilution of genetic integrity of a natural species by massive pollen flow from a crop leading to extinction of the natural species as it becomes more like the crop.

Genetic drift Changes in gene frequency in a population that arise through random events.

Genetic engineering Modifying the genetic makeup of living organisms using modern molecular biology techniques that can combine genes from widely dissimilar organisms.

Genome The genetic information contained in one complete set of chromosomes.

Genus Taxonomic category containing closely related species. Organisms in the same genus may occasionally interbreed.

Helper virus Virus with which satellite RNA is associated.

Hitchhiking A situation in which a neutral gene, because it is close on a chromosome (linked) to an advantageous crop gene, remains in a population because it is carried along with the advantageous gene.

Horizontal gene flow Movement of genes between adult individuals through nonsexual means.

Hybrid Offspring of two parents differing from each other in one or more genes.

Interfertile plants Plants that can interbreed and produce viable, fertile offspring.

Introgression Process by which new genes are introduced into a wild population by backcrossing of hybrids between two populations.

Invasiveness Capacity of a plant to spread beyond the site of introduction and become established in new sites.

Natural ecosystem Lightly managed or essentially unmanaged ecosystem, such as wilderness area, wildlife refuge, and habitat near fields.

Natural selection Process by which the interaction between organisms and the environment leads to a differential rate of reproduction among genetic types in a population. As a result, some genes increase in frequency in a population, while others decline. Natural selection is a primary factor in evolution.

Net replacement rate Measure of the capacity for a population to replace itself over time.

Nontarget organism Organism affected by a product even though it is not the intended recipient. For example, a pesticidal transgene product may affect organisms (nontarget ones) other than the pest at which it is directed.

Outbreeding depression Reduction in fitness that can follow hybridization.

Pathogen Organism capable of causing disease in another organism. Bacteria, fungi, and viruses are among the disease-causing agents of plants.

Persistence Capacity of a plant to remain in a setting for some period after it is introduced.

Phenology Study of phenomena, such as flowering, which occur periodically.

Phenotype Observable characteristics of an organism.

Plant-microbial interactions Interactions between plants and microorganisms, such as fungi and bacteria. Some of the interactions are temporary and casual; others are intimate and longer-term physical and functional associations. Interactions may occur on the surface of or internally in cells and tissues of leaves, stems, and roots. Examples of intimate associations include those between bacteria (rhizobia) and roots of legumes that result in nitrogen fixation.

Pleiotropy Effect of a gene on a number of different traits.

Pollination The transfer of pollen from the male to the female part of a flower.

Population replacement experiments Experiments described in this report to measure the rate at which a population multiplies itself and the persistence of its seed bank.

Propagules Structures produced by a plant that are capable of reproducing the plant. Sexual propagules are seeds; asexual propagules include rhizomes and runners.

Recombinant DNA DNA from different sources joined together by humans using modern molecular biology methods.

Recombination A natural process whereby an exchange of pieces of nucleic acid between two similar nucleic acids results in new combinations of genes; in plants, recombination typically occurs during sexual reproduction as chromosomes form new associations.

Replication Process by which copies of nucleic acid are made.

Reproductive success Capacity of a population to produce viable, fertile offspring.

Satellite RNA Small pieces of RNA that are sometimes associated with some strains of certain plant viruses; satellite RNA is not part of the genome of the virus with which it is associated but is dependent on the virus for replication.

Seed bank Repository of seeds in soil.

Self-pollination Pollen of one plant transferred to female part of same plant or another plant with the same genetic makeup.

Sexual reproduction Reproduction involving the union of female (egg) and male (pollen) cells. In flowering plants, seeds result from sexual reproduction.

Sexually compatible plants Two plants that are capable of reproduction by sexual means, that is, donating or receiving eggs or pollen (sperm) to produce new offspring.

Species A taxonomic category typically (but not always) containing organisms that actually or potentially may interbreed in nature. The name of a species consists of its genus and species designations, with the genus name placed first.

Synergism Interaction between viruses causing a more severe disease than either virus causes alone.

Systems-based agriculture Approaches to growing food and fiber that take into account relationships among the many components of a farm and integrate research and experience to solve problems in a whole-farm context.

Taxonomy A branch of science dealing with the classification of organisms.

Transcapsidation Partial or full enclosure of the nucleic acid of one virus by the coat protein of a different virus.

Transgene Gene from a dissimilar organism or an artificially constructed gene added by methods of molecular biology to another organism.

Transgenic plant Plant that has been genetically engineered using gene-splicing methods or plant that is the offspring of transgenic plants. Typically, a transgenic plant contains genetic material from at least one unrelated organism, for example, bacteria, viruses, animals, and other plants.

Vector Biotic agent that carries pollen or viruses from plant to plant. The major vectors for both pollen and viruses are insects.

Vertical gene flow Movement of genes from parent to offspring through sexual reproduction.

Viral genome Sum total of genetic material of a virus.

Virus Small particle that can reproduce only inside a living cell; typically composed of a nucleic acid core and a protein coat; generally causes disease in host organisms.

Weed Unwanted plant.

Weediness Any undesirable effect of a plant.

Sources: Curtis, 1983; Matthews, 1991; Rural Advancement Foundation International, 1992.

Notes

Chapter 1

1. Foreign genes spliced into plants by modern molecular techniques are called transgenes.

2. Conventional agriculture typically focuses on a single problem, for example, an insect pest on one crop grown on a farm, and tries to find a chemical or genetic solution to that one problem. By contrast, systems-based approaches consider synergistic and conflicting relationships among the many components of a farm—humans, crops, domestic animals, soil, weather, wildlife, predators, watershed—and try to integrate research and experience to solve problems in a whole-farm context.

3. Recent research highlights the unpredictability associated with gene splicing. Finnegan and McElroy (1994) summarized a number of studies showing that transgenes can be turned off after successful incorporation into the plant genome. The experiments suggest that plants can recognize and inactivate foreign genetic material. In some cases, a change in the environment can initiate the inactivation process, raising the possibility that some engineered traits may not be stable under ordinary field conditions.

Chapter 2

1. In late 1994, Upjohn sold Asgrow Seed Company to the Mexican company, Empresas La Moderna.

Chapter 3

1. Agricultural lands include farms and ranches; nonagricultural ecosystems include forests, meadows, grasslands, and wetlands. We do not address the special problems posed by home gardens.

2. This report uses the terms "wild/weedy" or "noncrop" to describe plants that are weeds in ecosystems affected by humans or wild plants that grow exclusively or predominantly in habitats not heavily influenced by humans.

3. There may be substantial delays between the arrival of a plant and its emergence as a weed. For example, weedy races of proso millet appeared in Canada only after more than two centuries of cultivation (Cavers and Bough, 1985).

4. In this report, scientific names are given for the less-known noncrop plants but not for well-known crops.

5. Williamson (1993) scored only nine of the twelve characters listed in table 3.2 because two apply only to species and one trait was not objectively measurable.

6. The calyx is the outer covering of a closed flower bud and can often be seen as small green structures at the base of the petals in an open flower.

7. Bt is the name given to insecticidal toxins produced by the soil bacterium *Bacillus thuringiensis*.

8. It should be noted, however, that once the hybrid is formed after the initial pollen flow, further gene movement may occur by dispersal of seed or vegetative propagules such as tubers or bulbs if the plant reproduces asexually (without combination of male and female sex cells).

9. Plant breeders list the female member of the cross first, the male second. An "x" is used between two members of a cross—in this case wild/weedy plant x crop—when the direction of pollen flow is important. Otherwise, a hyphen between the words denotes a hybrid (crop-wild/weedy plant). Therefore, a crop x weed hybrid is formed with the crop providing the egg and the weed the pollen. A weed x crop hybrid means that the weed provided the egg, and the crop provided the pollen. Two kinds of hybrid seed involving wild/weedy plants may be produced: crop x wild/weedy plant and wild/weedy plant x crop. Crop x wild/weedy plant hybrids are formed when pollen from wild/weedy relatives flows into the transgenic crop. This discussion, however, focuses primarily on the other type of hybrid: wild/weedy plant x crop hybrids, where pollen from the transgenic crop flows into populations of wild/weedy relatives.

10. This statement refers only to nuclear genes; that is, genes carried on chromosomes contained in the nuclei of egg and pollen. Plants also possess genetic material outside the nucleus of a cell—in structures called mitochondria and chloroplasts that reside in the cytoplasm surrounding the nucleus. This genetic material is not usually transferred in a pollen grain. The egg, however, contains this nonnuclear genetic material. Our discussion is generally confined to considerations of nuclear genes.

11. It should be noted that, despite the generally accepted notion that corn has no sexually compatible relatives in the United States, the crop reportedly hybridizes with a wild relative, gama grass (*Tripsacum dactyloides*), that is found in the United States (Goodman et al., 1987 and references therein). However, it remains unclear whether viable, fertile hybrids are produced under natural conditions (H. Wilson, personal communication, 1993).

12. In recognition of the uncertainty about the potential for genes to flow between crops and wild relatives in the United States, the USDA in 1994 initiated a research project to help identify which of North America's thirty most

important food crops are sexually compatible and coexist with wild relatives (U.S. Department of Agriculture, 1995).

13. Teosinte is a common name that applies to several different wild plants native to Mexico and Central America. Some populations are classified as a subspecies of the species *Zea mays* (which also includes corn). Others are classified as a different species of the genus *Zea*.

14. Alternatively, the hybrids may be more fit than either of their parents, a condition referred to as hybrid vigor. Whether hybridization will result in outbreeding depression or vigor cannot usually be predicted (R. Linder, personal communication, 1993).

15. Panelists at the workshop identified this issue as one they lacked the expertise to address. Because the panelists thought it was important, the authors prepared a separate discussion of this issue after the workshop and circulated it among experts in plant virology. Papers by de Zoeten (1991), Palukaitis (1991), and Tolin (1991) provided much of the information and many of the ideas and sources in this section.

16. Coat protein surrounds the nucleic acid core of plant viruses. The nucleic acid, most commonly RNA, contains genes that code for the coat protein and other proteins needed to produce new viral particles. Once engineers have spliced a coat protein gene into the plant's DNA, synthesis of the coat protein proceeds just as if it were a plant protein. The plant synthesizes coat protein messenger RNA (m-RNA) complementary to viral gene nucleic acid. That coat protein m-RNA then becomes the blueprint for synthesis of coat proteins on plant ribosomes.

17. Cross protection describes a phenomenon whereby a plant infected with a mild strain of a virus develops resistance to severe strains of the same virus. To obtain cross protection that is agronomically useful, susceptible crops are purposely infected with a mild strain of a virus. The crops subsequently develop resistance to damaging infection by severe strains of the same or similar viruses. The mechanism of cross protection is unknown (Matthews, 1991).

18. Although some viruses may be transmitted mechanically through sap, by vegetative propagation, and in seeds and pollen, the most common means of transmission is by insects (Matthews, 1991).

19. The phenomenon, also called heterologous encapsidation and genomic masking (Rochow, 1977), will be referred to as transcapsidation in this discussion.

20. Replication is the process by which copies of nucleic acid are made.

21. Encapsidation means that the RNA has been enclosed in coat protein (capsid).

22. Coat protein is a structural protein of a virus. Proteins produced by the virus that have other functions in the viral life cycle (for example, replication) are the nonstructural proteins.

23. Palukaitis (1991) describes two categories of coat protein-containing plants where transcapsidation is not expected to lead to expanded host range by insect

transmission: i) plants with coat protein of viruses not transmitted by insects and ii) plants with coat protein of viruses that require components in addition to coat protein for transmission.

24. Approximately one-fifth of known viruses are transmitted through seeds from infected plants. In some cases the virus is a contaminant on the surface of the seed. More commonly the virus infects the embryo within the seed. Seed transmission is common in some groups of viruses, rarely occurs in others, and in some groups has never been observed (Matthews, 1991).

25. This happens because the virus already inhabits the seeds and is likely to replicate in the new plants that grow from the seed in the next growing season.

26. For example, in some potatoes, synergism of potato viruses Y and X produces rugose mosaic (Matthews, 1981).

27. A defective virus is a form of virus which lacks the capacity to fulfill one or more critical biological functions such as replication, coat protein production, and assembly of viral particles.

Chapter 4

1. The ability to control a risk is often viewed as an element of risk management rather than assessment (for example, National Research Council, 1983). For simplicity, this book considers control as part of the assessment of risk.

2. Although the focus of this book is annual crops, nearly all of which reproduce sexually via seed, there may be situations where a risk assessment should consider the capacity for crops to reproduce asexually through vegetative spread or production of nonsexual propagules.

3. Seed bank refers to the reservoir of viable seeds remaining in the soil.

4. Genus (plural: genera) refers to the taxonomic classification that includes one or more species. Generally, plants in the same species are sufficiently similar that they can interbreed and produce viable offspring. For example, two kinds of corn that are sexually compatible and capable of producing viable hybrid offspring are generally placed in the same species *mays* of the genus *Zea*. Plants that share many similarities—but generally do not interbreed—are placed in the next level of relatedness: genus. For example, many wild plants are similar to the potato *Solanum tuberosum*, yet they do not readily interbreed. They are in the same genus, *Solanum*, but different species.

5. It should be kept in mind that varying definitions of "weed" will have been used to compile this information. Plants causing economic losses are far more likely to be reported as weeds than weeds that invade and alter wild ecosystems.

6. There is no comprehensive published flora of the United States or North America. More serious is the lack of published flora in tropical and subtropical areas where many centers of diversity exist (Doebley, 1988).

7. The USDA approved Asgrow's petition in late 1994.

8. Although this section relies on the nontransgenic crop as a point of comparison for the transgenic crop, it should not be interpreted as an acceptance of the commercialization of any nontransgenic crop. In fact, there may be existing and potential nontransgenic crops that present sufficient likelihood for adverse ecological consequences that they should not be planted widely.

9. Where plants reproduce asexually, the analysis compares the vegetative replacement capacity of the transgenic and nontransgenic plants. For example, data may be taken on the numbers of shoots or asexual propagules, or on the area covered by or weight of plant biomass.

10. Seed bank decline is measured in half-lives because of an assumed exponential decay rate. However, some recent work (Rees and Long, 1992) has shown that not all seed banks decay according to a negative exponential function. It may be necessary in particular cases to determine the most appropriate function to use in modeling seed decay rate.

11. Dormancy is a period in which germination does not occur; growth begins only when certain environmental requirements, for example, particular temperature and moisture levels, are met.

12. For example, suppose 1,000 seeds are planted the first year of the ecological performance trial. For simplicity, it is assumed that all 1,000 seeds germinate and produce a total of 2,000 seeds. If 90 percent of these 2,000 seeds become dormant, only 200 seeds will germinate and grow into plants that reproduce the second year. Assuming that these 200 plants produce 400 seeds and 90 percent are dormant, 40 seeds will sprout in the third year and produce 80 seeds. Except for the first year when 2,000 seeds were produced, it appears that the population is rapidly going to extinction. When the seed bank is taken into account, it becomes clear that this is not the case. Eighteen hundred seeds remain in the seed bank from the first year, 360 from the second year, and 72 from the third year, or a total of 2,232 seeds in the seed bank. Even if only half of these seeds survive to reproduce, there will have been an increase in the original population. (Example provided by R. Linder, personal communication, 1993.)

13. At one meeting, for example, some participants argued for a classification of risks of transgenes into three categories: (i) high-fitness genes that confer a broad defensive quality, such as insect or disease resistance and would persist a long time; (ii) moderate-fitness genes that confer a quality in a special setting, such as herbicide resistance; and (iii) low-fitness genes that would be disadvantageous and have low persistence, such as male sterility. In particular, participants felt that a cluster of traits, including weediness, as well as disease, insect, and herbicide resistance, deserved special attention (U.S. Department of Agriculture, 1990).

14. Discussions of relative performance assume that experimental data have been subject to statistical analyses and that differences in performance are based on statistically significant differences in net replacement rates and seed-bank half-lives.

15. $R_t \leq R_n$ and $H_t \leq H_n$, where t = transgenic and n = nontransgenic.

16. $R_t < 1$.

17. $R_t > R_n$ across all environments, provided $R_t > 1$.

18. One additional consideration is that the capacity to form hybrids differs among varieties of a crop. Unless the sexual compatibility is consistent across varieties of the crops, the tier 1 analysis cannot be eliminated.

19. For example, a wild relative of corn in the United States, gama grass (*Tripsacum dactyloides*) can hybridize with corn (Goodman et al., 1987 and references therein) but it remains unclear whether the two plants are interfertile under natural conditions (H. Wilson, personal communication, 1993).

20. An alternative concept that might be used for regulatory purposes is the gene pool system (Hancock, 1990 and references therein). This approach recognizes varying levels of interfertility between a crop and its wild relatives (Hancock, 1990). The primary gene pool (GP-1) of a crop includes relatives that are fully crossable with the crop, resulting in generally fertile hybrids. In a crop's secondary gene pool (GP-2) are relatives with which hybridization is possible, but difficult, and where hybrids are weak and have reduced fertility. Finally, the tertiary gene pool (GP-3) contains relatives that may form hybrids with the crop, but the hybrids are not viable or are sterile. Using this system for regulatory purposes, one might suggest that relatives residing in the United States and part of the primary and secondary gene pools of a crop be subject to testing for interfertility with the crop.

21. The margin of safety is not absolute. A few crops may interbreed with plants that share higher taxonomic designations. This is the case among many cacti and orchids (N. Ellstrand, personal communication, 1992). Wheat (*Triticum aestivum*), for example, is interfertile with a weed, jointed goat grass, placed by some taxonomists in a different genus (*Aegilops cylindrica*) (Bowden, 1959; Donald and Ogg, 1991). In these cases, the studies of relatives must be conducted at the family or higher level.

22. In addition to the wild/weedy plant x crop hybrids, other hybrids may have formed. They are hybrids between the transgenic crop and nearby nonengineered crops, crop x crop (for example, transgenic corn will cross-pollinate nearby nonengineered corn) and crop x wild/weedy plant hybrids formed from pollen flowing from relatives into the transgenic crop. Crop x crop and crop x wild/weedy hybrids will be harvested and dispersed with the crop. These hybrids may warrant assessment but their evaluation is beyond the scope of this book.

23. Ultimately, models predicting gene flow and their establishment in wild/weedy populations may substitute for some portion of the empirical data now needed to predict a transgene's impact in wild/weedy populations (Kareiva, 1990; Kareiva, et al., 1991; Manasse and Kareiva, 1991).

24. Replacement rate ideally should measure the number of offspring of male and female parents. The approach suggested in this book measures only the maternal component, that is, the parent that contributed the egg to the cross. Seed production is a measure of the maternal contribution. Not measured is the paternal component, that is, the offspring formed from pollen contributed by the transgenic hybrid to other plants in the population. In contrast to seed produc-

tion as a relatively simple measure of maternity, no practical means are currently available for paternity analyses. Therefore, in the near future, assessments will rely on maternity analyses (R. Linder, personal communication, 1993).

25. F_1 hybrid refers to the result of a cross between a transgenic crop and a wild plant; F_2 refers to the offspring of interbreeding between F_1 hybrids.

26. Backcrossing refers to interbreeding of the F_1 wild/weedy plant x crop hybrids with wild/weedy plants. Backcrossing can go on for many generations, with the genetic content of each succeeding generation becoming more and more similar to that of the unhybridized wild/weedy plants.

27. It should be noted that results indicating that transgenic and nontransgenic wild/weedy plants perform similarly mean that the transgene is likely to be retained in the population, even though it gives no selective advantage (R. Linder, personal communication, 1993).

28. About one-fifth of the applications to test transgenic crops in the United States since 1987 have been for virus-resistant crops (Union of Concerned Scientists, 1994).

Chapter 5

1. Corn is generally considered such an "ecological cripple" that it is a good candidate for reduced testing in tier 2 under our scheme and perhaps, after some greater accumulation of field experience, for a categorical exemption from field testing requirements.

2. A relative of corn, gama grass (*Tripsacum dactyloides*) is found in the United States. While it can be made to hybridize with corn experimentally (Goodman et al., 1987 and references therein), it remains unclear whether it hybridizes under natural conditions (H. Wilson, personal communication, 1993).

3. There is controversy about the origin of corn. Some scientists believe that rather than corn descending from teosinte, both corn and teosinte are the descendants of a common ancestor, now extinct (Beadle, 1980).

4. Swamping of small isolated populations of wild plants could also occur with nonengineered plants containing disadvantageous traits. Genetically engineered cultivars might pose a greater risk in this regard because of the potentially destabilizing properties of novel genes discussed in chapter 3.

5. Mexico has a program that monitors and seeks to protect its annual populations of teosinte. It also has established a biosphere reserve to protect the only known stands of a perennial teosinte (*Zea diploperennis*), a primitive wild relative of corn (National Research Council, 1993).

Appendix

1. Blocks represent different field margins in different environments (growing areas with different environmental conditions).

2. Allow interactions among plots during growing season.

3. Where crops are engineered to resist stresses, such as herbicides, insects, diseases, and temperature extremes, no additional application of stress beyond that found naturally over the range of environments tested should be applied.

4. Instead of counting all seeds, weigh a known subset, count the number of seeds in the subset, weigh all seeds, and calculate the total number.

5. Viability of seeds can be measured directly by germinating a subset, and, for those that do not germinate, indirectly using a dye-uptake method (Delouche et al., 1962).

6. Genetic type determination may require molecular markers and identification methods, such as polymerase chain reaction (PCR) methods (Innis et al., 1989).

References

Alabama Cooperative Extension Service. No date. *Kudzu in Alabama: history, uses, and control.* Circular ANR-65. Auburn University.

Allison, R., C. Thompson, and P. Ahlquist. 1990. Regeneration of a functional RNA genome by recombination between deletion mutants and requirement for cowpea chlorotic mottle virus 3a and coat genes for systemic infection. *Proceedings National Academy of Sciences* 87:1820–24.

AMBIO. 1992. The economics of biodiversity loss. Special issue, vol. 21, no. 3.

Andow, D., P. Kareiva, S. Levin, and A. Okubo. 1990. Spread of invading organisms. *Landscape Ecology* 4:177–88.

Asgrow Seed Company. 1992. Petition for determination of regulatory status. Submitted to U.S. Department of Agriculture Animal and Plant Health Inspection Service, July 13.

Aspelin, A., A. Grube, and R. Torla. 1992. Pesticides industry sales and usage: 1990 and 1991 market estimates. Washington, D.C.: Environmental Protection Agency, Office of Pesticide Programs, Economic Analysis Branch.

Baker, H. 1965. Characteristics and modes of origin of weeds. In *The Genetics of Colonizing Species*, ed. H. G. Baker and G. L. Stebbins, 147–68. New York: Academic Press.

Baker, H. 1972. Migration of weeds. In *Taxonomy, Phytogeography, and Evolution*, ed. D. Valentine, 327–47. London: Academic Press.

Baker H. 1974. The evolution of weeds. *Annual Review of Ecology and Systematics* 5:1–24.

Barrett, S. 1983. Crop mimicry in weeds. *Economic Botany* 37:255–82.

Baskin, J., and C. Baskin. 1985. The annual dormancy cycle in buried weed seeds: a continuum. *BioScience* 35:492–98.

Beachy, R. 1990. Coat-protein mediated protection against virus infection in transgenic plants. In *Risk Assessment in Agricultural Biotechnology: Proceedings of the International Conference*, J. J. Marois and G. Bruening, technical authors, 21–28. Davis: University of California.

Beachy, R., S. Loesch-Fries, and N. Tumer. 1990. Coat protein-mediated resistance against virus infection. *Annual Review of Phytopathology* 28:451–74.

Beadle, G. 1980. The ancestry of corn. *Scientific American* 242:112–19.

Biotech Patent News. 1992. Calgene Fresh enters into production agreements with leading fresh tomato packers. Vol. 6, no. 5, p. 5, August.

Biotechnology Industry Organization (BIO). 1995. *Agricultural Biotechnology: The Future of the World's Food Supply*. Washington, D.C.

Bowden, W. 1959. The taxonomy and nomenclature of the wheats, barleys, and ryes and their wild relatives. *Canadian Journal of Botany* 37:657–84.

Boyce Thompson Institute. 1987. *Regulatory Considerations: Genetically Engineered Plants*. San Francisco: Center for Science Information.

Brown, L., and J. Young. 1990. Feeding the world in the nineties. In *State of the World, 1990*, 59–78. New York: W. W. Norton.

Brunken, J., J. de Wet, and J. Harlan. 1977. The morphology and domestication of pearl millet. *Economic Botany* 31:163–74.

Bujarski, J., and P. Kaesberg. 1986. Genetic recombination between RNA components of a multipartite plant virus. *Nature* 321:528–31.

Burdon, J., R. Groves, and J. Cullen. 1981. The impact of biological control on the distribution and abundance of *Chondrilla juncea* in southeastern Australia. *Journal of Applied Ecology* 18:957–66.

Callihan, R., D. Swan, D. Thill, and D. Morishita. 1990. Jointed goatgrass: a threat to wheat. University of Idaho, College of Agriculture Cooperative Extension System, Miscellaneous Series 141. North Idaho Foundation Seed Association, Inc., Pullman, Wash.

Caplan, A., and M. Van Montagu. 1990. Evolutionary consequences of modifying cultivated plants. In *Introduction of Genetically Modified Organisms Into the Environment*, ed. H. A. Mooney and G. Bernardi, 57-68. Scientific Committee on Problems of the Environment 44, Chichester, N.Y.: John Wiley and Sons.

Carlson, T., and B. Chelm. 1986. Apparent eukaryotic origin of glutamine synthetase II from the bacterium *Bradyrhizobium japonicum*. *Nature* 322:568–70.

Cavers, P., and M. Bough. 1985. Proso millet (*Panicum miliaceum* L.): a crop and a weed. In *Studies on Plant Demography: A Festschrift for John L. Harper*, ed. J. White, 143–55. London: Academic Pres.

Chen, B., and R. Francki. 1990. Cucumovirus transmission by the aphid *Myzus persicae* is determined solely by the viral coat protein. *Journal of General Virology* 71:939–44.

Collmer, C., and S. Howell. 1992. Role of satellite RNA in the expression of symptoms caused by plant viruses. *Annual Review of Phytopathology* 30:419–42.

Cook, R. 1980. The biology of seeds in the soil. In *Demography and Evolution in Plant Populations*, ed. O. Solbrig, 107–29, Botanical Monographs Number 15. Berkeley: University of California Press.

Council for Agricultural Science and Technology. 1991. *Herbicide-Resistant Crops*. Ames, Iowa.

Crawley, M. 1990a. The ecology of genetically engineered organisms: assessing the environmental risks. In *Introduction of Genetically Modified Organisms Into the Environment*, ed. H. A. Mooney and G. Bernardi, 133–50. Scientific Committee on Problems of the Environment 44, Chichester, N.Y.: John Wiley and Sons.

Crawley, M. 1990b. PROSAMO Plant Programme, Imperial College, Experimental Results for 1990. Department of Biology, Imperial College, Silwood Park, Ascot, Berks, SL5 7PY, United Kingdom.

Crawley, M. 1992. The comparative ecology of transgenic and conventional crops. In *The Biosafety Results of Field Tests of Genetically Modified Plants and Microorganisms*, ed. R. Casper and J. Landsmann, 43–52. Biologische Bundensanstalt für Land- und Forstwirtschaft, Braunschweig.

Crawley, M., R. Hails, M. Rees, D. Kohn, and J. Buxton. 1993. Ecology of transgenic oilseed rape in natural habitats. *Nature* 363:620–23.

Curtis, H. 1983. *Biology*. New York: Worth Publishers, Inc.

Dale, P. 1994. The impact of hybrids between genetically modified crop plants and their related species: general considerations. *Molecular Ecology* 3:31–36.

Darmency, H. 1994. The impact of hybrids between genetically modified crop plants and their related species: introgression and weediness. *Molecular Ecology* 3:37–40.

Day, P. 1974. *Genetics of Host-Parasite Interaction*. San Francisco: W. H. Freeman.

Decker, D. S. 1988. Origin(s), evolution, and systematics of *Cucurbita pepo* (Cucurbitaceae). *Economic Botany* 42:4–15.

Decker-Walters, D. S. 1990. Evidence for multiple domestication of *Cucurbita pepo*. In *Biology and Utilization of the Cucurbitaceae*, ed. D. A. Bate, et al., 96–101. Ithaca, N.Y.: Cornell University Press.

Delouche, J., T. Still, M. Raspet, and M. Lienhard. 1962. The tetrazolium test for seed viability. Mississippi State University Agricultural Experiment Station Technical Bulletin 51:1–63.

de Zoeten, G. 1991. Risk assessment: do we let history repeat itself? *Phytopathology* 81:585–86.

Doebley, J. 1988. Environmental issues, panel 5. In *Proceedings of a USDA/EPA/FDA Transgenic Plant Conference*, September 7–9, 1988, 137–145. Annapolis, Md.

Doebley, J. 1990. Molecular evidence for gene flow among *Zea* species. *BioScience* 40:443–48.

Donald, W., and A. Ogg. 1991. Biology and control of jointed goatgrass (*Aegilops cylindrica*), a review. *Weed Technology* 5:3–17.

Doyle, J. 1985. *Altered Harvest*. New York: Viking.

Duffey, S., and K. Bloem. 1986. Plant defense-herbivore-parasite interactions and biological control. In *Ecological Theory and Integrated Pest Management Practice*, ed. M. Kogan, 135–84. New York: John Wiley and Sons.

Economist. 1995. Science and technology: plant workers. May 16, 79–80.

Ehrlich, P., and G. Daily. 1993. Population extinction and saving biodiversity. *AMBIO* 22:64–68.

Ehrlich, P., and A. Ehrlich. 1992. The value of biodiversity. *AMBIO* 21:219–26.

Ellstrand, N. 1988. Pollen as a vehicle for the escape of engineered gene? *Trends in Ecology and Evolution* 3:S30–S32, *Trends in Biotechnology* 6:S30–S32 (combined issue).

Ellstrand, N. 1992. Gene flow by pollen: implications for plant conservation genetics. *Oikos* 63:77–86.

Ellstrand, N., B. Devlin, and D. Marshall. 1989. Gene flow by pollen into small populations: data from experimental and natural stands of wild radish. *Proceedings of the National Academy of Sciences* 86:9044–47.

Ellstrand, N., and C. Hoffman. 1990. Hybridization as an avenue of escape for engineered genes. *BioScience* 40:438–42.

Ellstrand, N., and D. Marshall. 1985. Interpopulation gene flow by pollen in wild radish, *Raphanus sativus*. *American Naturalist* 126:606–16.

Environmental Protection Agency. 1992. Fifth developmental toxicity peer review of bromoxynil. Memo from G. J. Burin and A. Clevenger to J. McQueen. Office of Pesticide Programs, April 21.

Environmental Protection Agency. 1995. Bromoxynil: pesticide tolerance. *Federal Register* 60:16111–14.

Eshbaugh, W. 1976. XII. Genetic and biochemical systematic studies of chili peppers (*Capsicum*—Solanaceae). *Bulletin of the Torrey Botanical Club* 102:396–403.

Ewel, J. 1986. Invasibility: lessons from South Florida. In *Ecology of Biological Invasions of North America and Hawaii*, ed. H. A. Mooney and J. A. Drake, 214–30. New York: Springer-Verlag.

Falk, B., and G. Breuning. 1994. Will transgenic crops generate new viruses and new diseases? *Science* 263:1395–96.

Farinelli, L., P. Malnoe, and G. Collet. 1992. Heterologous encapsidation of potato virus Y strain O (PVY^{O}) with the transgenic coat protein of PVY strain N (PVY^{N}) in *Solanum tuberosum* cv. Bintje. *Bio/Technology* 10:1020–25.

Finnegan, J., and D. McElroy. 1994. Transgene inactivation: plants fight back! *Bio/Technology* 12:883–88.

Flexner, J., B. Lighthart, and B. Croft. 1986. The effects of microbial pesticides on non-target, beneficial arthropods. *Agriculture, Ecosystems and Environment* 16:203–54.

Fowler, C., and P. Mooney. 1990. *Shattering: Food, Politics, and the Loss of Genetic Diversity*. Tucson: University of Arizona Press.

Foy, C., D. Forney, and W. Cooley. 1983. History of weed introductions. In *Exotic Plant Pests and North American Agriculture*, ed. C. L. Wilson and C. L. Graham, 65–92. New York: Academic Press.

Fraley, R. 1992. Sustaining the food supply. *Bio/Technology* 10:40–43.

Friends of the Earth European Coordination. 1993. Evaluation of biosafety information gathered during field releases of GMO's [DSTI/STP/BS (92)6]. *Clearinghouse on Biotechnology*, Mail-out no. 16, 7–11.

Fuller, T., and E. McClintock. 1986. *Poisonous Plants of California*. California Natural History Guides: 53, Berkeley: University of California Press.

Gal, S., B. Pisan, T. Hohn, N. Grimsley, and B. Hohn. 1992. Agroinfection of transgenic plants leads to viable cauliflower mosaic virus by intermolecular recombination. *Virology* 187:525–33.

Gershon, D. 1992. Programme aids developing world. *Nature* 356:735.

Gliddon, C. 1994. The impact of hybrids between genetically modified crop plants and their related species: biological models and theoretical perspectives. *Molecular Ecology* 3:41–44.

Golemboski, D., G. Lomonossoff, and M. Zaitlin. 1990. Plants transformed with a tobacco mosaic virus nonstructural gene sequence are resistant to the virus. *Proceedings National Academy of Sciences* 87:6311–15.

Goodman, R., H. Hauptli, A. Crossway, and V. Knauf. 1987. Gene transfer in crop improvement. *Science* 236:48–54.

Gordon, K., and R. Symons. 1983. Satellite RNA of cucumber mosaic virus forms a secondary structure with partial 3'-terminal homology to genomal RNAs. *Nucleic Acids Research* 11:947–60.

Gould, F., and R. Shaw. 1968. *Grass Systematics*. College Station: Texas A & M University Press.

Graham, F. 1970. *Since Silent Spring*. New York: Houghton Mifflin.

Granatstein, D. 1988. *Reshaping the Bottom Line: On-Farm Strategies for a Sustainable Agriculture*. Stillwater, Minn.: Land Stewardship Project.

Grant, V. 1981. *Plant Speciation*. Second edition. New York: Columbia University Press.

Greene, A., and R. Allison. 1994. Recombination between viral RNA and transgenic plant transcripts. *Science* 263:1423–25.

Hancock, J. 1990. *Plant Evolution and the Origin of Crop Species*. Englewood Cliffs, N.J.: Prentice Hall.

Harlan, J. 1982. Relationships between weeds and crops. In *Biology and Ecology of Weeds*, ed. W. Holzner and N. Numata, 91–96. The Hague: W. Junk Publishers.

Harrison, B., M. Mayo, and D. Baulcombe. 1987. Virus resistance in transgenic plants that express cucumber mosaic virus satellite RNA. *Nature* 328:799–802.

Harrison, S., L. Oliver, and D. Bell. 1977. Control of Texas gourd in soybeans. *Proceedings Southern Weed Science Society* 30:46.

Hauptli, H., N. Newell, and R. Goodman. 1985. Genetically engineered plants: environmental issues. *Bio/Technology* 3:437–42.

Hawkes, J. 1983. *The Diversity of Crop Plants*. Cambridge, Mass.: Harvard University Press.

Heinemann, J. 1991. Genetics of gene transfer between species. *Trends in Genetics* 7:181–85.

Hirsch, A. 1992. Developmental biology of legume nodulation. *New Phytologist* 122:211–37.

Hoffman, C. 1990. Ecological risks of genetic engineering of crop plants. *BioScience* 40:434–37.

Hunn, J., E. Multer, and M. DeFelice. 1989. Fish and agricultural chemicals: safeguarding your pond. Columbia: University of Missouri Extension Division.

Iltis, H., and J. Doebley. 1980. Taxonomy of *Zea* (Gramineae) II: Subspecific categories in the *Zea mays* complex and a generic synopsis. *American Journal of Botany* 67:994–1004.

Innis, M., D. Gelfand, J. Sninsky, and T. White. 1989. *PCR Protocols: A Guide to Methods and Applications*. San Diego: Academic Press.

Jain, S., and P. Martins. 1979. Ecological genetics of the colonizing ability of rose clover (*Trifolium hirtum* All.). *American Journal of Botany* 66:361–66.

Juma, C. 1989. *The Gene Hunters: Biotechnology and the Scramble for Seeds*. Princeton: Princeton University Press.

Kareiva, P. 1990. Using models of population spread to analyze the results of field release. In *Risk Assessment in Agricultural Biotechnology: Proceedings of the International Conference*, J. J. Marois and G. Bruening, technical authors, 168–74. Davis: University of California.

Kareiva, P., R. Manasse, and W. Morris. 1991. Using models to integrate data from field trials and estimate risks of gene escape and gene spread. In *Biological Monitoring of Genetically Engineered Plants and Microbes*, ed. D. R. MacKenzie and S. C. Henry, 31–42. Bethesda, Md.: Agricultural Research Institute.

Keeler, K. 1989. Can genetically engineered crops become weeds? *Bio/Technology* 7:1134–39.

Keeler, K., and C. Turner. 1991. Management of transgenic plants in the environment. In *Risk Assessment in Genetic Engineering: Environmental Release of Organisms*, ed. M. Levin and H. Strauss, 189–218. New York: McGraw-Hill, Inc.

Kennedy, G. 1986. Plant-plant-pathogen-insect interactions. In *Ecological Theory and Integrated Pest Management Practice*, ed. M. Kogan, 203–16. New York: John Wiley and Sons.

Kirkpatrick, K., and H. Wilson. 1988. Interspecific gene flow in *Cucurbita*: *C. texana* and *C. pepo*. *American Journal of Botany* 75:519–27.

Klinger, T., P. Arriola, and N. Ellstrand. 1992. Crop-weed hybridization in radish (*Raphanus sativus*): effects of distance and population size. *American Journal of Botany* 79:1431–35.

Kogan, M. 1986. Plant defense strategies and host-plant resistance. In *Ecological Theory and Integrated Pest Management Practice*, ed. M. Kogan, 83–134. New York: John Wiley and Sons.

Krimsky, S., J. Ennis, and R. Weissman. 1991. Academic-corporate ties in biotechnology: a quantitative study. *Science, Technology, and Human Values* 16:275–87.

Kurath, G., and P. Palukaitis. 1990. Serial passage of infectious transcripts of a cucumber mosaic virus satellite RNA clone results in sequence heterogeneity. *Virology* 176: 8–15.

Langevin, S., K. Clay, and J. Grace. 1990. The incidence and effects of hybridization between cultivated rice and its related weed red rice (*Oryza sativa* L.) *Evolution* 44:1000–1008.

Lecoq, H., M. Ravelondandro, C. Wipf-Scheibel, M. Monsion, B. Raccah, and J. Dunez. 1993. Aphid transmission of a non-aphid-transmissible strain of zucchini yellow mosaic potyvirus from transgenic plants expressing the capsid protein of plum pox potyvirus. *Molecular Plant-Microbe Interactions* 6:403–406.

Lenski, R. 1988a. Experimental studies of pleiotropy and epistasis in *Escherichia coli* I: Variation in competitive fitness among mutants resistant to virus T4. *Evolution* 42:425–32.

Lenski, R. 1988b. Experimental studies of pleiotropy and epistatis in *Escherichia coli* II: Compensation for maladaptive effects associated with resistance to T4. *Evolution* 42:433–40.

Levin, D. 1978. The origin of isolating mechanisms in flowering plants. In *Evolutionary Biology*, ed. M. Hecht, W. Steere, and B. Wallace, 185–317. New York: Plenum Press.

Levings, C. 1990. The Texas cytoplasm of maize: cytoplasmic male sterility and disease susceptibility. *Science* 250:942–47.

Lewin, R. 1982. Can genes jump between eukaryotic species? *Science* 217:42–43.

Linder, C., and J. Schmitt. 1994. Assessing the risks of transgene escape through time and crop-wild hybrid persistence. *Molecular Ecology* 3:23–30.

Lommel, S., and Z. Xiong. 1991. Reconstitution of a functional red clover necrotic mosaic virus by recombinational rescue of the cell-to-cell movement gene expressed in a transgenic plant. *Journal of Cellular Biochemistry* 15A:151 (abstr).

Mack, R. 1988. Environmental issues, panel 5. In *Proceedings of a USDA/EPA/FDA Transgenic Plant Conference*, September 7–9, 1988, 127–35. Annapolis, Md.

Manasse, R. 1992. Ecological risks of transgenic plants: effects of spatial dispersion on gene flow. *Ecological Applications* 2:431–38.

Manasse, R., and P. Kareiva. 1991. Quantifying the spread of recombinant genes and organisms. In *Assessing Ecological Risks of Biotechnology*, ed. L. R. Ginzburg, 215–31. Boston: Butterworth-Heinemann.

Manasse, R., and K. Pinney. 1991. Limits to reproductive success in a partially self-incompatible herb: fecundity depression at serial life-cycle stages. *Evolution* 45:712–20.

Mann, C., and M. Plummer. 1992. The butterfly problem. *The Atlantic Monthly*, 47–59, January.

Matthews, R. 1981. *Plant Virology*. New York: Academic Press.

Matthews, R. 1991. *Plant Virology*. New York:Academic Press.

McGarvey, P., J. Kaper, M. Avila-Rincon, L. Pena, and J. Diaz-Ruiz. 1990. Transformed tomato plants express a satellite RNA of cucumber mosaic virus and produce lethal necrosis upon infection with viral RNA. *Biochemical and Biophysical Research Communications* 170:548–55.

McGaughey, W., and M. Whalon. 1992. Managing insect resistance to *Bacillus thuringiensis* toxins. *Science* 258:1451–55.

McGranahan, G., J. Hansen, and D. Shaw. 1988. Inter- and intraspecific variation in California black walnuts. *Journal American Society Horticultural Science* 113:760–65.

Mellon, M., and J. Rissler. 1995. Transgenic crops: USDA data on small-scale tests contribute little to commercial risk assessment. *Bio/Technology* 13:96.

Miller, J. 1990. Field assessment of the effects of a microbial pest control agent on nontarget Lepidoptera. *American Entomologist*, 135–39, summer 1990.

Moffat, A. 1995. Exploring transgenic plants as a new vaccine source. *Science* 268:658, 659.

Molnar, R., and G. Ingram. 1991. Reticulating the tree of life. *Australian Natural History* 23:736–37.

Mooney, H., S. Hamburg, and J. Drake. 1986. The invasions of plants and animals into California. In *Ecology of Biological Invasions of North America and Hawaii*, ed. H. A. Mooney and J. A. Drake, 250–72. New York: Springer-Verlag.

National Research Council. 1983. *Risk Assessment in the Federal Government: Managing the Process*. Washington, D.C.: National Academy Press.

National Research Council. 1989a. *Alternative Agriculture*. Washington, D.C.: National Academy Press.

National Research Council. 1989b. *Testing Genetically Modified Organisms: Framework for Decisions*. Washington, D.C.: National Academy Press.

National Research Council. 1993. *Managing Global Genetic Resources: Agricultural Crop Issues and Policies*. Washington, D.C.: National Academy Press.

Nee, M. 1990. Domestication of *Cucurbita* (Cucurbitaceae). *Economic Botany* 44 (3 Supplement):56–68.

Ninio, J. 1983. *Molecular Approaches to Evolution*. Princeton, N.J.: Princeton University Press.

Oliver, L., S. Harrison, and M. McClelland. 1983. Germination of Texas gourd (*Cucurbita texana*) and its control in soybeans (*Glycine max*). *Weed Science* 31:700–06.

Osbourn, J., S. Sakar, and M. Wilson. 1990. Complementation of coat protein-defective TMV mutants in transgenic tobacco plants expressing TMV coat protein. *Virology* 179:921–25.

Palm, C., K. Donegan, D. Harris, and R. Seidler. 1994. Quantification in soil of *Bacillus thuringiensis* var. *kurstaki* delta-endotoxin from transgenic plants. *Molecular Ecology* 3:145–51.

Palukaitis, P. 1991. Virus-mediated genetic transfer in plants. In *Risk Assessment in Genetic Engineering: Environmental Release of Organisms*, ed. M. Levin and H. Strauss, 140–62. New York: McGraw-Hill, Inc.

Panestos, C., and H. Baker. 1967. The origin of variation in "wild" *Raphanus sativus* (Cruciferae) in California. *Genetica* 38:243–74.

Perrins, J., M. Williamson, and A. Fitter. 1992. A survey of differing views of weed classification: implications for regulation of introductions. *Biological Conservation* 60:47–74.

Perry, D., M. Amaranthus, J. Borchers, S. Borchers, and R. Brainerd. 1989. Boot-strapping in ecosystems. *BioScience* 39:230–37.

Pirone, T. 1991. Viral genes and gene products that determine insect transmissibility. *Seminars in Virology* 2:81–87.

Powell-Abel, P., R. Nelson, B. De, N. Hoffmann, S. Rogers, R. Fraley, and R. Beachy. 1986. Delay of disease development in transgenic plants that express the tobacco mosaic virus coat protein gene. *Science* 232:738–43.

Rees, M., D. Kohn, R. Hails, M. Crawley, and S. Malcolm. 1991. An ecological perspective to risk assessment. In *Biological Monitoring of Genetically Engineered Plants and Microbes*, ed. D. R. MacKenzie and S. C. Henry, 9–24. Bethesda, Md.: Agricultural Research Institute.

Rees, M., and M. Long. 1992. Germination biology and the ecology of annual plants. *American Naturalist* 139:484–508.

Regal, P. 1982. Pollination by wind and animals: ecology of geographic patterns. *Annual Review of Ecology and Systematics* 13:497–524.

Regal, P. 1986. Models of genetically engineered organisms and their ecological impact. In *Ecology of Biological Invasions of North America and Hawaii,* ed. H. A. Mooney and J. A. Drake, 111–29. New York: Springer-Verlag.

Regal, P. 1988. The adaptive potential of genetically engineered organisms in nature. *Trends in Ecology and Evolution* 3:S36–S38, *Trends in Biotechnology* 6:S36–S38 (combined issue).

Regal, P. 1990. Gene flow and adaptability in transgenic agricultural organisms: long-term risks and overview. In *Risk Assessment in Agricultural Biotechnology: Proceedings of the International Conference*, J. J. Marois and G. Bruening, technical authors, 102–10. Davis: University of California.

Regal, P. 1992. Scientific principles for ecologically based risk assessment of transgenic plants. Draft prepared for the symposium Potential Ecological and Nontarget Effects of Transgenic Plant Gene Products on Agriculture, Silviculture, and Natural Ecosystems, November 30–December 2, 1992, College Park, Md.

Robbins, W., M. Bellue, and W. Ball. 1970. *Weeds of California.* State of California, Sacramento.

Rochow, W. 1977. Dependent virus transmission from mixed infections. In *Aphids As Virus Vectors,* ed. K. F. Harris and K. Maramorosch, 253–73. New York: Academic Press.

Rural Advancement Foundation International. 1992. The seed map: dinner on the Third World. Pittsboro, N.C.: RAFI-USA.

Ryan, J. 1992. Conserving biological diversity. In *State of the World, 1992,* 9–26. New York: W. W. Norton.

Sanford, J., T. Klein, E. Wolf, and N. Allen. 1987. Delivery of substances into cells and tissues using a particle bombardment process. *Particulate Science and Technology* 5:27–37.

Scheffler, J., R. Parkinson, and P. Dale. 1993. Frequency and distance of pollen dispersal from transgenic oilseed rape (*Brassica napus*). *Transgenic Research* 2:356–64.

Schoelz, J., K-B. Goldberg, and J. Kiernan. 1991. Expression of cauliflower mosaic virus (CaMV) gene VI in transgenic *Nicotiana bigelovii* complements a strain of CaMV defective in long-distance movement in nontransformed *N. bigelovii. Molecular Plant-Microbe Interactions* 4:350–55.

Schoelz, J., and W. Wintermantel. 1993. Expansion of viral host range through complementation and recombination in transgenic plants. *The Plant Cell* 5:1669–79.

Seed World. 1992. Seed giants. 26–33, November.

Shatters, R., and M. Kahn. 1989. Glutamine synthetase II in *Rhizobium*: reexamination of the proposed horizontal transfer of DNA from eukaryotes to prokaryotes. *Journal of Molecular Evolution* 29:422–28.

Simmonds, N. 1976. *Evolution of Crop Plants.* New York: Longman.

Sleat, D., and P. Palukaitis. 1990. Site-directed mutagenesis of a plant viral satellite RNA changes its phenotype from ameliorative to necrogenic. *The Plant Journal* 2:43–49.

Sleat, D., and P. Palukaitis. 1992. A single nucleotide change within a plant virus satellite RNA alters the host specificity of disease induction. *The Plant Journal* 2:43–49.

Slobodchikoff, C., and J. Doyen. 1977. Effects of *Ammophila arenaria* on sand dune arthropod communities. *Ecology* 58:1171–75.

Small, E. 1984. Hybridization in the domesticated-weed-wild complex. In *Plant Biosystematics*, ed. W. F. Grant, 195–210. Toronto: Academic Press.

Smid, D., and L. Hiller. 1981. Phytotoxicity and translocation of glyphosate in the potato (*Solanum tuberosum*) prior to tuber initiation. *Weed Science* 29:218–23.

Smith, B., C. Cowan, and M. Hoffman. 1992. Is it an indigene or a foreigner? In *Rivers of Change*, B. Smith, 67–100. Washington, D.C.: Smithsonian Institution Press.

Templeton, A. 1986. Coadaptation and outbreeding depression. In *Conservation Biology: The Science of Scarcity and Diversity*, ed. M. Soule, 105–16. Sunderland, Mass.: Sinauer Assoc.

Tepfer, M. 1993. Viral genes and transgenic plants: what are the potential environmental risks? *Bio/Technology* 11:1125–30.

Thomas, H., and J. Mytton. 1970. Mono-somic analysis of fatuoids in cultivated oat *Avena sativa. Canadian Journal of Genetics and Cytology* 12:32–35.

Thomas, P., and D. Smith. 1983. Relationship between cultural practices and the occurrence of volunteer potatoes in the Columbia Basin. *American Potato Journal* 60:289–94.

Thompson, D., R. Stuckey, and E. Thompson. 1987. Spread, impact and control of purple loosestrife (*Lythrum salicaria*) in North American wetlands. Fish and Wildlife Research 2. U.S. Department of the Interior, Fish and Wildlife Service, Washington, D.C.

Tien, P. 1990. Satellite RNA for the control of plant diseases. In *Risk Assessment in Agricultural Biotechnology: Proceedings of the International Conference*, J.J. Marois and G. Bruening, technical authors, 29–37. Davis: University of California.

Tien, P., and G. Wu. 1991. Satellite RNA for the biocontrol of plant disease. *Advances in Virus Research* 39:321–39.

Till-Bottraud, I., X. Reboud, P. Brabant, M. Lefranc, B. Rherissi, F. Vedel, and H. Darmency. 1992. Outcrossing and hybridization in wild and cultivated foxtail millets: consequences for the release of transgenic crops. *Theoretical and Applied Genetics* 83:940–46.

Tolin, S. 1991. Persistence, establishment, and mitigation of phytopathogenic viruses. In *Risk Assessment in Genetic Engineering: Environmental Release of Organisms*, ed. M. Levin and H. Strauss, 114–39. New York: McGraw-Hill, Inc.

Turner, C. 1988. Ecology of invasions by weeds. In *Weed Management in Agroecosystems: Ecological Approaches*, ed. M. A. Altieri and M. Liebman, 41–55. Boca Raton, Fla.: CRC Press.

Union of Concerned Scientists. 1993. Compilation of data from applications to the U.S. Department of Agriculture to field test transgenic crops. Washington, D.C.

Union of Concerned Scientists. 1994. Compilation of data from applications to the U.S. Department of Agriculture to field test transgenic crops. Washington, D.C.

United Nations Industrial Development Organization. 1992. International biosafety guidelines and code of conduct for the release of genetically engineered microorganisms and plants. *Genetic Engineering and Biotechnology Monitor* 39:1–19, September.

U.S. Congress Office of Technology Assessment. 1993. *Harmful Non-Indigenous Species in the United States*. OTA-F-566. Washington, D.C.

U.S. Department of Agriculture. 1987. Introduction of organisms and products altered through genetic engineering which are plant pests or which there is reason to believe are plant pests. *Federal Register* 52:22892–915.

U.S. Department of Agriculture. 1990. *Proceedings: Workshop on Safeguards for Planned Introduction of Transgenic Oilseed Crucifers*. Animal and Plant Health Inspection Service. October 9, 1990, Cornell University, Ithaca, N.Y.

U.S. Department of Agriculture. 1992. Scientific evaluation of the potential for pest resistance to the *Bacillus thuringiensis* (Bt) delta-endotoxins. Cooperative State Research Service, Agricultural Research Service, Conference to Explore Resistance Management Strategies, Washington, D.C.

U.S. Department of Agriculture. 1993. Genetically engineered organisms and products; notification procedures for the introduction of certain regulated articles; and petition for nonregulated status. *Federal Register* 58:17044–59.

U.S. Department of Agriculture. 1995. Risk assessment research: an overview of the USDA Biotechnology Risk Assessment Research Grants Program. Washington, D.C.: Cooperative State Research, Education, and Extension Service and Agricultural Research Service.

Vitousek, P. 1986. Biological invasions and ecosystem properties: can species make a difference? In *Ecology of Biological Invasions of North America and Hawaii*, ed. H. A. Mooney and J. A. Drake, 163–76. New York: Springer-Verlag.

Waterworth, H., J. Kaper, and M. Tousignant. 1979. CARNA 5, the small cucumber mosaic virus-dependent replicating RNA, regulates disease expression. *Science* 204:845–47.

Weed Science Society of America. 1989. *Herbicide Handbook*. Herbicide Handbook Committee, Champaign, Ill.

Wilkes, H. 1977. Hybridization of maize and teosinte, in Mexico and Guatemala and the improvement of maize. *Economic Botany* 31:254–93.

Williams, M. 1980. Purposefully introduced plants that have become noxious or poisonous weeds. *Weed Science* 28:300–305.

Williamson, M. 1992. Environmental risks from the release of genetically modified organism(GMOs)—the need for molecular ecology. *Molecular Ecology* 1:3–8.

Williamson, M. 1993. Invaders, weeds and the risk from genetically manipulated organisms. *Experientia* 49:219–224.

Williamson, M. 1994. Community response to transgenic plant release: predictions from British experience of invasive plants and feral crop plants. *Molecular Ecology* 3:75–79.

Wilson, E. 1992. *The Diversity of Life*. Cambridge, Mass.: Harvard University Press.

Wilson, H. 1990. Gene flow in squash species. *BioScience* 40:449–55.

Wilson, H. 1993. Free-living *Cucurbita pepo* in the United States: viral resistance, gene flow, and risk assessment. Order #43-6395-3-C4203, prepared for USDA Animal and Plant Health Inspection Service, Hyattsville, Md.

Wisconsin Rural Development Center. 1993. Private interests, public responsibilities and the College of Agriculture and Life Sciences. Mount Horeb, Wis.

Young, J. 1990. Bred for the hungry? *World Watch* 3(1):14–22.

Index